CIVIL SERVICE COLLEGE
MANAGEMENT IN GOVERNMENT

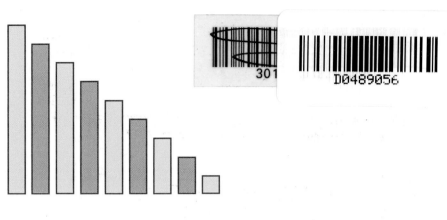

PLAIN FIGUReS

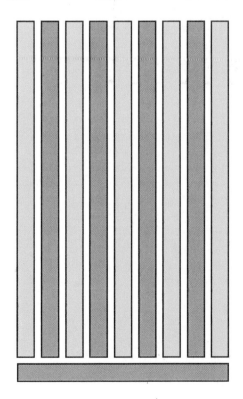

SECOND EDITION BY
MYRA CHAPMAN AND CATHY WYKES

London: The Stationery Office

ISBN 0 11 702039 7

f 001.4224

f 519.5

Contents

Acknowledgements

Our thanks are due to all the friends and colleagues in the Government Statistical Service who, when they heard that Plain Figures was to be updated, offered constructive suggestions and provided examples of good and bad practice in presenting numbers. They also gave us copies of departmental codes of practice in drawing up statistical charts and tables. Particular thanks are due to Alison Holding and Ed Swires Hennessy for their detailed comments on our first complete draft of this edition and to Peter Stibbard for his positive support and the helpful suggestions he offered throughout the project. We are grateful to the editors of many Government statistical publications for permission to use their tables and charts: particularly useful were Social Trends, and many publications from the, then, Employment Department. Our task would have been infinitely harder if we had been unable to use these publications as rich stores of good (and not so good) examples of statistical tables and charts.

A humble tribute is also due to Sir Ernest Gowers, whose *Complete Plain Words* was the original stimulus to writing *Plain Figures*. We hope that our book will advance the cause of clear communication a littler further by encouraging a similarly disciplined, thoughtful yet simple approach to the representation of numerical information.

The Stationery Office wishes to acknowledge the following for their kind permission to reproduce copyright material: British Telecom for the table 'Analysis of shareholders' taken from *BT Directors report and financial statement 1993* © British Telecommunications plc; the Office for Official Publications of the European Communities for the tables 'The European climate', 'Goods transport in the European Community 1989 (per cent)', 'Major EC seaports and main inland waterways' and 'Lung cancer mortality, 1986' from *Europe in Figures* and 'European agriculture in the world: a major force' from *European File. A community of twelve: key figures. Eurostat*; the *Financial Times* of 9 March 1994 for the figure 'Social Security: the claims on state spending'; the *New Scientist* of 23 October 1993 for the figure 'Unnatural shrinkage: since the 1960s over 70 per cent of the Aral Sea's water has been lost'; The British Council for the figure extracted from the *British Council annual report and accounts 1991/92*; *Autocar and Motor* of 26 January 1994 for the figures 'Cars on the road at end of 1992 by year of registration', 'Features you'd change', 'Taxation of company cars' and 'Getting away with it'; The Bank of England for the figure 'Type of transaction – principals' turnover' from the *Bank of England Quarterly Bulletin* November 1992 and 'Personal sector saving and borrowing' from *Bank of England Inflation Report* February 1994; *The Economist* for the figure 'Going, going, gone' © *The Economist*, London, 26 June 1993; The British Petroleum Company plc. for the figure 'World energy consumption per capita' from *BP Statistical Review of World Energy* June 1994; the *Independent* of 24 March 1994 for the figure 'Britain's changing shopping basket'; the US Bureau of the Census, Washington DC, for the figure 'Changes in farm populations, 1940 to 1973' from *Statistical Abstract of the United States: 1974* (95th edition); and the SERPLAN *Regional Monitoring Commentary* for the figure 'Housing provision in South East England 1981–2001'.

If the legal copyright holders of material not acknowledged above wish to contact The Stationery Office, we shall ensure that the correct acknowledgement appears in any future edition of this book.

Introduction

Aim of this book

Most people are not statisticians. But most people are regularly confronted with figures: in newspaper articles on unemployment or inflation, in reports on the performance of companies or products, in advertisements, in trade union claims for equity. Wherever the reader is invited to make a quantitative comparison, the writer thinks (correctly) that the comparison will be clearer if figures are quoted.

But clarity is not always achieved and phrases like 'Lies, damned lies, and statistics', 'Of course you can prove anything with statistics', and 'I've never had a head for figures' have become common defensive reactions. This is a great pity. Properly used, numbers provide by far the most effective way of describing changes and making comparisons in all the above areas and in many others.

The aim of this book is to demonstrate and discuss ways of presenting numbers effectively so that their value can be realised in full. A subsidiary aim is to help the reader interpret all data more competently and confidently.

It is important to state immediately that an honest intention is assumed. This book has no handy hints for those who wish to disguise the shortcomings of their data or who wish to twist their data to support a pre-conceived argument.

Who is it for?

The book is intended for anyone who needs to present statistical information, including:

- statisticians
- economists
- scientists and specialists in other numerate disciplines
- administrators, managers and others who sometimes statistical information in the course of their work.

It is written for those who wish to communicate a quantitative message to a non-specialist audience. No statistical techniques are discussed. The book deals with ways of communicating numbers effectively without using terms like 'standard deviation' and 'level of significance'. Statisticians, economists and scientists will often use formal statistical techniques to analyse data but, once they

have identified the main patterns in the data, the guidelines in this book may help them to present their findings to a non-technical audience.

Although as a Civil Service College publication the book is designed primarily to meet the needs of civil servants, it naturally covers the needs of others also, since the principles of good statistical presentation hold for all applications.

Why is it needed?

Statistical presentation is fundamentally important: if it is not done properly all prior work on data collection and analysis is likely to be wasted. Despite this, presentation is seldom taught in formal academic courses and has few generally known principles or standards. Consequently it is frequently done badly through lack of method, discipline, or thought about its purpose.

Badly presented statistical information can mislead people and may lead to mistakes in further calculations: it may waste people's time, and no matter how relevant, is likely to be ignored. Conversely, well presented information can be assimilated quickly and accurately and can therefore be used with ease and confidence.

For theses reasons a book is needed which will encourage people to think more about their objectives in presenting quantitative information, and which will provide useful principles and guidelines on effective methods of presentation.

This book brings together advice and research findings on statistical presentation which have appeared in a variety of sources: from articles published in the 1920s to work on tables and graphs done by Professor A S C Ehrenberg in the 1970s and 1980s.

Plan of the book

The book starts with a chapter on the general principles of statistical presentation, differentiating clearly between data presented for reference purposes and data presented to illustrate an argument.

The second chapter contains a summary of the arguments and recommendations developed throughout the book. This chapter is included to provide a summary of the principles of good data presentation for those with limited time, and to direct such readers to chapters which may be of immediate personal interest. It also provides those who have read the whole book with a quick reminder of the important guidelines developed in later chapters.

The next three chapters are devoted to tables: first a chapter on the structure common to all statistical tables, then a chapter on tables designed to communicate specific messages, and thirdly a chapter on tables designed for reference purposes.

Chapters 6 and 7 are about the use of pictures, or charts, to illustrate quantitative messages, Chapter 6 dealing with general principles and Chapter 7 discussing some specific kinds of statistical chart which can be used effectively in presenting quantitative results to a non-specialist audience. Finally the role of narrative in communicating statistical information is discussed in Chapter 8.

The book contains a great many tables and charts, some demonstrating good practice, some exhibiting various faults. For ease of reference, good tables and charts have been identified by the symbol ☑

The research findings which support the guidelines offered for effective presentation of data are reviewed in Appendix C. This appendix is included to assure the reader that the guidelines discussed are not just based on personal opinions which may safely be disregarded. The guidelines stem from a variety of sources: experimental evidence, the considered opinions of experienced practitioners, evidence from the field of cognitive psychology - and, occasionally, personal opinion. Since the book offers considerably more positive advice than usual on which medium to use for which message, care has been taken to explain the basis of its recommendations.

It will be obvious to any reader who knows the work of Professor Ehrenberg how much this book owes to his work on tables and graphs.

Sources of data

Many tables and charts in this book have been reproduced directly from other publications and these are listed in Appendix E. The tables and charts usually include a source reference: a survey, a government department, another publication, etc. This is the source of the original data. For example Table 4.3 includes a reference to the Digest of UK Energy Statistics as the data source, however the table has in fact been reproduced from the Digest of Environmental Protection and Water Statistics. Our comments on the table relate to the table design in the Environment publication as shown in appendix E.

The energy publication is the source of data, should readers decide to seek more information

The second edition

The first edition of *Plain Figures* was out of date almost as soon as it was published in 1986 - its worthy advice on producing good tables using a standard typewriter having been overtaken by the widespread use of wordprocessors and graphics packages. These tools allow the writer of a statistical report to produce a wide variety of thoroughly professional tables and charts. They also allow writers to produce elaborate and gimmicky muddles. The second edition of Plain Figures, just like the first, offers guidance on producing effective tables and charts, by whatever means the writer has available. We have assumed that those who produce statistical charts and tables use wordprocessors, spreadsheets and graphics packages - and are likely to produce the final version of a chart or table themselves. For this reason, the second edition offers rather more advice than the first on details which affect the final appearance (and therefore effectiveness) of a chart or table.

Statistical publications, and their tables and charts, appear in colour - sometimes used to good effect, sometimes not. However, since we have been unable to find any authoritative research on the effect of using different colours or different colour combinations in quantitative presentations, guidance on the use of colour is confined to a number of simple, commonsense observations. These are at the end of Chapter 3. The charts and tables have been scanned in, so that the colour and print quality is as close as possible to the original, but may sometimes be slightly poorer. They provide examples of a wide range of colours and colour combinations and studying which tables and charts have made best use of colour will help readers to use colour to good effect in their own work.

1 General principles for presenting data

1.1 Reasons for presenting data

A set of data may appear in a report or publication for two quite distinct reasons. It may be presented for future reference, that is, to provide a range of numerical information from which other people will select the data they need for specific analyses. Or it may be presented to demonstrate a particular fact or to support an argument.

The principles to be applied in presenting data for future reference are different from those to be applied when presenting data for demonstration purposes. This is because reference data and demonstration data are produced at different stages in the process of collecting and using statistical data and are produced for different reasons. The collection and use of statistical data can be divided into the following stages:

a   collect and check the basis data
b   collate and store the data in a convenient form for later use
c   select and analyse an appropriate subset of the data
d   report findings
e   influence decisions.

The final stage is of course the ultimate purpose of collecting data: if the data cannot be used to influence decisions, the cost and effort of earlier stages are wasted.

Reference data are produced at stage **b** and may be stored in computer files or recorded in tables, in the expectation that they will be used in later analyses. Many government statistical publications are primarily data storage devices which are produced at this stage. Demonstration data, however, are produced at stage **d**, generally in the expectation that they will be used at stage **e**. Different criteria must therefore be applied to the presentation of each sort of data: data designed for future reference must be presented for ease of use; data intended to support an argument or demonstrate a particular fact must be presented for ease of comprehension.

1.2 Selection of medium for presenting statistics

There are only three ways of presenting statistical information: it can appear in a table, in the form of a picture (a chart, a graph or a diagram of some sort) or in a paragraph of prose. These three media - tables, pictures and words - may be used in the following ways:

	Reference	Demonstration
Tables	Yes	Yes
Pictures	No	Yes
Words	No	Yes

If figures are being presented for *reference* purposes there is usually no sensible choice but tables. Reading numbers from a graph or chart tends to be slow and inaccurate, and words alone would be an exceedingly inefficient means of recording reference data.

For *demonstration* purposes the most effective combination of tables, pictures and words should be chosen. And here it is important to use the media in mutually supporting ways. Tables and charts very seldom 'speak for themselves' so if either is to be effective in demonstrating a particular point it should be accompanied by a verbal summary. Both tables and pictures should be clear and simple and should be included in the main text rather than in an annex at the back.

In general, when choosing how to present demonstration data, tables are best for conveying numerical values, pictures are best for conveying qualitative relationships and words are best for conveying implications for action.

For example, Figure 1.1 and Table 1.1 both communicate the following messages: first that three quarters of freight is carried by road, far more than by any other mode of transport: second that there was a decrease in the proportion of goods carried by rail between 1955 and 1990, but increases in the proportions carried by road, water and pipeline: and third that the total amount of freight transported increased by about 60 per cent (from 1,300 million tonnes to 2,200 million).

The table is more likely to communicate the actual numbers effectively: the totals of 1,300 in 1955 and 2,200 in 1990, and the proportion of freight not going by road which fell from 25 per cent in 1955 (21+3+1) to 19 per cent in 1990 (6+7+6).

Table 1.1 Freight transport by mode, 1955 and 1990

Great Britain		Percentages
	1955	1990
Road	75	81
Rail	21	6
Water	3	7
Pipeline	1	6
Total (100%) (million tonnes)	1,300	2,200

Source: Department of Transport

Figure 1.1 Freight transport by mode, 1955 and 1990

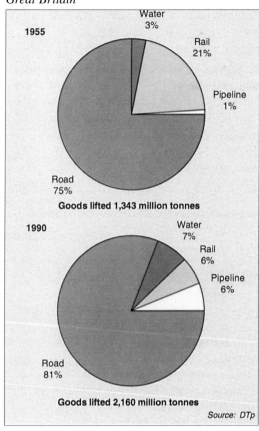

Great Britain

1955
Water 3%
Rail 21%
Pipeline 1%
Road 75%
Goods lifted 1,343 million tonnes

1990
Water 7%
Rail 6%
Pipeline 6%
Road 81%
Goods lifted 2,160 million tonnes

Source: DTp

The chart immediately highlights two qualitative relationships: the dominance of road transport in both years and the shift away from rail transport to other modes.

It is usually more expensive to produce a good chart than a good table, both in terms of design time and, particularly when colour is used, the cost of reproduction. However, charts are more decorative and are likely to appeal more to a non-specialist audience. In choosing between tables and charts, a sensible rule is to use a mix of charts and tables. For a numerate, specialist audience more tables and fewer charts can be used, but for a non-specialist audience the balance shifts toward charts.

1.3 Tables and charts for demonstration Does it work?

In considering the merits of any table or chart designed for demonstration purposes, only one question needs to be asked, 'Does it work?' In other words, does the table or chart communicate a message to the reader clearly and simply?

Before discussing the general principles of communicating numbers effectively, it is helpful to examine examples of good and bad practice. The rest of section 1.3 is devoted to examining some examples of demonstration tables and charts. At first reading you may prefer to skip straight to section 1.4, returning to consider these examples later.

Table 1.2 Composition of consumer credit

United Kingdom	Percentages and £ billion			
	1982	1987	1991	1992
Bank credit card lending	12.7	16.7	18.2	18.7
Bank loans[1]	66.5	62.7	62.8	62.6
Finance houses[2]	8.3	11.9	10.2	9.5
Insurance companies	2.0	2.5	2.8	2.9
Retailers	10.5	6.0	4.7	4.9
Building Society loans[3]	0.0	0.2	1.3	1.4
Credit outstanding at end of year (= 100%)(£ billion)	15.9	36.2	53.8	52.9

1 Banks and all other institutions authorised to take deposits under the *Banking Act 1987.*
2 Finance houses and other credit companies (excluding institutions authorised to take deposits under the *Banking Act 1987*).
3 Building Society unsecured loans to individuals or companies (i.e. Class 3 loans as defined in the *Building Societies Act 1986*).

Source: Central Statistical Office

Verbal summary to Table 1.2

Bank loans still remain the main source of consumer credit in the United Kingdom, although their overall share continues to decline to stand at just under 63 per cent in 1992. Bank credit card lending, on the other hand, continues to rise steadily accounting for 13 per cent of outstanding credit in 1982 and 19 per cent in 1992. Retailers accounted for only five per cent of consumer credit in 1992, half the proportion of ten years earlier.

Let us start by considering Tables 1.2 and 1.3. The first table is reproduced from *Social Trends 24* along with its verbal summary. The same data are presented rather differently in Table 1.3.

Table 1.2 is attractively laid out, clearly labelled and accompanied by a verbal commentary. But most people will find the layout of Table 1.3 more helpful.

There are a number of differences between the two tables, and, in each case, Table 1.2 has been altered to make it easier for the reader to assimilate the patterns in the data.

First, all entries in the second table have been rounded: this allows readers to carry out mental arithmetic easily and hence to identify trends and exceptions to general trends: we can all subtract 63 from 67 and get 4, but more effort is involved in subtracting 62.7 from 66.5. It is rarely desirable to include a decimal place in a table showing percentage shares.

Secondly, the rows and columns have been interchanged. The reason for this is that the main patterns in the table are trends over time and it is

Table 1.3 Composition of consumer credit

United Kingdom	Credit source (percentages)						Percentages and £ billion
	Bank loans[1]	Bank credit card lending	Finance houses[2]	Retailers	Insurance companies	Building society loans[3]	Credit outstanding at year end (£ billion)
1982	67	13	8	10	2	..	16
1987	63	17	12	6	2	..	36
1991	63	18	10	5	3	1	54
1992	63	19	9	5	3	1	53

[1] Banks and all other institutions authorised to take deposits under the Banking Act 1987.
[2] Finance houses and other credit companies (excluding institutions authorised to take deposits under the Banking Act 1987).
[3] Building Society unsecured loans to individuals or companies (i.e. Class 3 loans as defined in the Building Societies Act 1986).

Source: Central Statistical Office

Verbal Summary to Table 1.3

There has been a more than threefold increase in consumer credit (excluding house purchase) from 1982 to 1992 - from £16 billion at the end of 1982 to £53 billion at the end of 1992. Adjusting these figures for inflation, consumer credit has still doubled over the period. Within this total....

easier to compare figures in columns than figures in rows. The sharp increase in the proportion of bank credit card lending from 13 to 17 per cent between 1982 and 1987 followed by a slower increase to 19 per cent in 1992 is obvious when the figures are arranged vertically:

13
17
18
19

but less apparent when the numbers are arranged horizontally:

13   17   18   19

Two reasons contribute to this: numbers in columns are physically closer to each other than numbers in rows and, as in the case of drastic rounding, this layout helps the reader to carry out mental arithmetic. At school, subtraction sums were taught by putting one number on top of another and the answer underneath; few people were trained to do horizontal sums in their formative years.

A third difference is that the order of the categories has been changed so that the largest single category in 1992 (bank loans) appears first, then the next largest and so on across the page. The reason for ordering columns like this is to remove arbitrary irregularity from the table. If the reader sees immediately that, in general, numbers decrease from left to right (or, in other cases, from top to bottom) any exceptions to this pattern are identified quickly and are easier to remember as exceptions to a general trend.

The verbal summary accompanying Table 1.2 starts with the clearest message in the data - the continuing dominance of bank loans as a source of consumer credit - and goes on to note three other changes, all clearly substantiated in the table. This is a helpful and workmanlike summary which reinforces the messages in the table and confirms to the reader that he or she has not missed a key feature of the data. (There is another notable feature in Table 1.2, namely the more than three fold growth in the value of credit outstanding at the year end. This, however, is dealt with in a separate chart and commentary in *Social Trends 24*. Were the table to be freestanding, the verbal summary should start by noting this feature - as has been done in the verbal summary of Table 1.3.)

Chapter 8 (section 8.6) summarises good practice in writing a verbal summary.

Just as a good table makes the patterns and exceptions obvious at a glance, so a good chart instantly communicates a clear message to the reader. By contrast a poor chart leaves the reader puzzling over its interpretation.

The essential difference between Figure 1.2 and 1.3 is that one has a clear story to tell and the other has not. In Figure 1.2 a great deal of data from the Youth Cohort Study has been carefully converted into graphical form, but what pattern emerges? Because the activities of the young people are shown cumulatively it is very difficult to identify differences between cohorts in any single activity (other than "school" which is at the bottom of the bar). Instead of helpfully directing the reader's attention to any memorable patterns in the data, the commentary states baldly:

> YCS can show what is happening in the youth and labour market for different age groups and over time.

This leaves the onus on the reader to work out what exactly *is* shown and what are the most significant changes and comparisons.

By contrast Figure 1.3 is much less ambitious yet much more effective. Comparatively few data are shown, but clear patterns are immediately obvious and these are highlighted in the accompanying text. The chart has a number of messages to communicate:

- the total amount of lowland grassland has decreased by almost 40 per cent since 1932 (from almost 8 million hectares to under 5 million hectares)
- the area of semi-natural grassland has declined to around 0.2 million hectares, compared with 5.8 million in 1932
- the area of rough grazing has also reduced significantly, by 70 per cent over the same period
- the area devoted to leys and improved grassland has increased.

In Figure 1.3, the chart and commentary work together to communicate information to the reader clearly and memorably. They would work together even better if the terms used in the text related directly to the units used in the charts (that is percentages were related to millions of hectares) and the text referred simply to "lowland grassland" (as in the chart) rather than "lowland grassland not intensified for agriculture". The text also includes additional explanations which help the lay reader understand the terms and changes more clearly.

Figure 1.2 Activities of young people, 16-17 years old

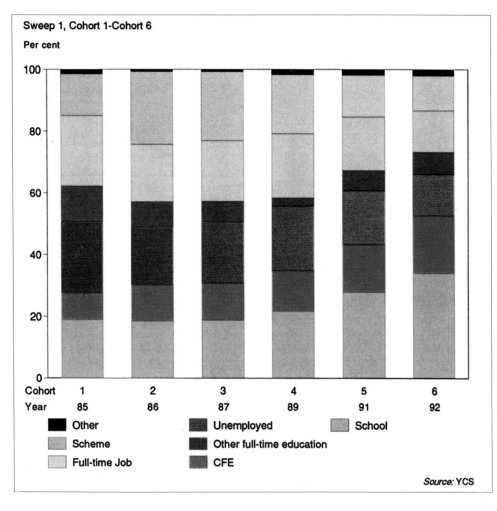

Sweep 1, Cohort 1-Cohort 6
Per cent

Cohort	1	2	3	4	5	6
Year	85	86	87	89	91	92

■ Other   ■ Unemployed   ■ School

■ Scheme   ■ Other full-time education

■ Full-time Job   ■ CFE

Source: YCS

Verbal summary to Figure 1.2

YCS can show what is happening in the youth labour market for different age groups and over time. Because it is a time series, comparisons can be made between different cohorts. For instance, YCS has clearly shown the change in staying on rates in education, the subsequent changes in labour market participation, and the effect of fluctuations in general economic conditions.

Figure 1.3 Lowland grassland, 1932 to 1984

England and Wales

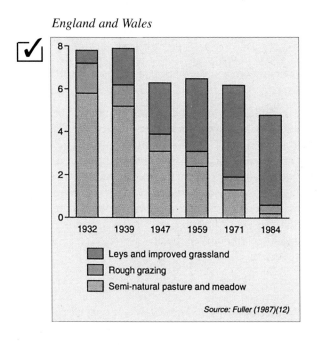

| 1932 | 1939 | 1947 | 1959 | 1971 | 1984 |

■ Leys and improved grassland

■ Rough grazing

■ Semi-natural pasture and meadow

Source: Fuller (1987)(12)

Verbal summary to Figure 1.3

In England and Wales, the amount of lowland grassland not intensified for agriculture, has decreased by almost 40 per cent since the first comprehensive survey in 1932. The area of semi-natural grassland has declined to around 0.2 million hectares compared with 5.8 million in 1932. Some has been converted into agriculturally improved grassland by application of herbicides and fertilizers and re-seeding, but most of the area has been ploughed and replaced by arable crops and grass leys. The area of rough grazing on commons and upland fringes has also reduced significantly, by 70 per cent over the same period.

5

1.4 General principles for demonstration tables and charts

The design of demonstration tables is discussed in detail in Chapter 4, and Chapters 6 & 7 are devoted to graphical presentation. In both cases, the prime objective is to devise a method of communicating patterns in the original data as effectively as possible.

In the case of tables this involves displaying numbers in a way which allows readers to compare numbers with each other as easily as possible (using mental arithmetic if necessary) and in a way which emphasises any general patterns in the data (such as a decrease in magnitude from left to right across the table or from the top to the bottom of the table): exceptions to the general pattern are then readily identifiable.

In the case of charts the same general principle leads to the use of simple, familiar forms of chart, like pie charts, bar charts of various sorts, and line graphs showing how one or more measurements varied over time. It also limits the amount of information that can be encapsulated effectively in a single chart to two or three clear patterns.

Both tables and charts should always be accompanied by a verbal summary which repeats, and therefore reinforces, the message illustrated. This summary should be close enough to the table or chart for readers to glance back from the summary to the table or chart to check their understanding of each point mentioned: the "Ah yes, I see" reaction should be encouraged.

1.5 Reference tables Are they easy to use?

The design of reference tables is discussed in detail in Chapter 5. The objective is simply to provide the data required in a form that the reader will find easy to use. This involves giving as much detail as can practicably be provided and including helpful totals and subtotals. It will also frequently involve choosing categories and an order of tabulation which are consistent with related tables. It is important to include complete definitions of the terms used in the table and to note any changes in definition of terms or in methods of data collection which may effect the validity of comparisons between different entries in the tables.

1.6 Accuracy

In reference tables it is important to record the intrinsic accuracy of the data. This can be done explicitly by giving ranges or standard errors (if appropriate to the expected user) or implicitly by rounding all figures appropriately.

In demonstration tables, most figures should be presented in rounded form. Inevitably this involves some loss of accuracy, but it is important to realise, in this context, that if an apparent pattern in the data vanishes when figures are rounded appropriately, it is doubtful if the pattern is robust enough to form a basis for decision-making. This point is discussed in more detail in Chapter 4.

1.7 Summary of general principles

Reference tables should be easy to use speedily and accurately. Demonstration tables and charts should communicate a small number of messages clearly and simply: they always need an accompanying verbal summary.

2 Summary of recommendations

2.1 Overview

This chapter presents, in condensed form, the guidelines which are discussed and illustrated in the rest of the book. It may be read alone but, to understand the reasons behind each recommendation and to see the effect of following specific guidelines, the appropriate chapter should be read in full.

2.2 Choice of medium for presentation

The first question to be resolved in presenting numbers is whether to use words, charts or tables. Few people would argue in favour of using words alone to present more than two or three isolated numbers, and so the usual choice is between tables and charts.

Most books and articles on the subject of presenting data fall into one of two categories: they either assume a commitment to graphical presentation and contain details of many ingenious graphical devices, or else they adopt a non-committed standpoint on whether tables or charts should be used to present statistical data.

The introductory chapter promised some positive guidance and this chapter summarises the recommendations discussed in the rest of the book. If these guidelines are followed carefully, even the most inexperienced presenter should produce effective tables and charts.

2.3 Demonstration or reference?

When planning to display any set of data, the first question to ask is

> Are these data being presented to demonstrate a particular point or are they for future reference?

If the answer is "for future reference", a table should be used, and a summary of the guidelines for constructing reference tables will be found in Section 2.9. Otherwise, read on. (Please try to avoid saying "Both": no chart or table can serve both purposes well. This point is elaborated in Chapter 3.)

2.4 For demonstration, charts or tables? Write the verbal summary first

Good charts tend to be time-consuming and expensive to produce, so it is important to know when a chart is likely to illustrate the patterns in a set of data more effectively than a table. Table 2.1 on page 10 lists the main advantages and disadvantages of charts. However an initial assessment of whether or not a chart is likely to be more effective than a table can be made by considering what sort of patterns the data exhibit.

Since neither charts nor tables ever "speak for themselves", in order to communicate a message effectively they must be accompanied by a verbal summary. And the final choice of whether to use a chart or a table is best made *after* writing the verbal summary.

The summary should be a clear, concise statement of how the data contribute to the subject of the report, consisting of no more than three or four separate points each clearly supported by the accompanying table or chart. Chapter 8 offers guidance on how to write a verbal summary of a table or chart. Before deciding on a verbal summary, a number of tables and rough graphs will probably have been constructed and the words finally chosen will often suggest the appropriate form of data display. On occasions both a table and a graph may be needed.

2.5 Tables for specific amounts

If the verbal summary emphasises *specific amounts*, then a table should be used. For example, the following is an extract from a verbal summary:

The increase in the number of Soviet military advisers in the Third World countries from 3,700 in 1965 (in 14 such countries) to about 15,000 in 1982 (in 31 such countries) is less well-known but ...

This clause stresses both the number of Soviet military advisers and the number of Third World countries in which they are based. Here the summary calls for an accompanying table which repeats the numbers quoted in the text and shows corresponding numbers for some intermediate years.

2.6 Charts for comparisons

Alternatively, in the verbal summary emphasises broad *comparisons*, such as:

> In social classes C1, C2, D and E, the *Sun* is the most popular daily newspaper; in social classes A and B the *Daily Telegraph* is most popular...

or if it includes a number of *non-specific* quantitative statements, such as "the trend levelled off" or "the percentage of families owning a colour television set increased dramatically", then a chart is likely to be more effective. Charts do, however, require more space and they can often take more time to prepare. Section 2.11 contains a more detailed analysis of what sort of chart is likely to be most appropriate in specific circumstances.

2.7 Which numbers to present

Demonstration tables may contain either the original data, often heavily rounded for ease of comprehension, or derived statistics such as percentages, indices or ratios. The choice of which numbers to present in the final table is best made after analysing the data in a number of different ways and establishing the main patterns and exceptions. The numbers which best illustrate the patterns and exceptions are the ones to present. For example, if the changes in the composition of a total are emphasised, *percentages* should usually appear in the final table. If the growth patterns in several different measurements are to be compared over a number of years, comparisons may be easier to make if all the measurements are expressed as *indices*, based on a common year whose value is taken as 100 in each of the series being examined. If two related measurements (such as total population and national spending on health) have grown at different rates, it may be informative to tabulate their *ratio* (£s per head). If an annual total results from an increase (such as births) and a decrease (deaths), it may be appropriate to tabulate the *differences* between these measurements as a net annual gain or loss. In some cases, valid comparisons can only be made if the original numbers of observations are re-expressed as *rates per number at risk* (for example, births in different areas would be related to the number of women of child-bearing age; accidents in different sports might be related to "player hours" or "player sessions"). And growth in any financial measurement can usually be assessed correctly only after it has been expressed *at constant prices*.

2.8 Guidelines for demonstration tables

The criterion for a good demonstration table is that "the patterns and exceptions should be obvious at a glance, at least once one knows what they are". To this aim the following basic guidelines are proposed:

a   Round numbers to two effective digits: this facilitates mental arithmetic and makes the numbers easier to remember.

b   Put the numbers most likely to be compared with each other in columns rather than rows: like guideline **a**, this makes quick comparisons easier since the numbers are physically closer to each other in columns than in rows.

c   Arrange columns and rows, where possible, in some natural order of size: if there is no obvious natural order, arrange columns and/or rows in decreasing order of their averages. This helps to impose as much visible structure as possible on the table, and exceptions to a general decreasing pattern are immediately highlighted. A corollary to guideline **c** is that wherever possible large numbers should appear at the top of columns, again to aid mental arithmetic (we subtract by putting the larger number on top), and to highlight the average patterns in the data. But an exception to this guideline must be made in government statistical publications where *time* trends are displayed in columns: in virtually all official statistical tables the most recent time period is shown at the *bottom* of columns, this practice should be followed for readers accustomed to this layout.

d   Give row and/or column averages or totals as a focus: they provide a structure which helps the reader to see the "average" relationships between rows and/or columns, and this in turn makes it easier to note and remember deviations from the average pattern.

e   Use layout to guide the eye: columns and rows in the body of the table should be regularly spaced and close together; additional blank spaces may be used to separate rows or columns showing averages or totals from the rest of the table; if the table continues more than six rows, an artificial break should be used to divide columns into blocks of four or five items for ease of reading (but demonstration tables should seldom be big enough to require such an artificial break).

f Accompany each table with a clear verbal summary, positioned close to the table and referring directly to numbers which appear in the table: this allows the reader to check his understanding of how the table supports the text, thus reinforcing the message and making it more memorable. No more than three or four distinct points should be made and these should summarise the *main patterns* in the data. Any necessary explanations of how the data were collected or collated should be shown separately.

A final guideline in constructing demonstration tables is that they should be small. It is better to include three or four compact tables, each illustrating one or two points succinctly, rather than to construct a single large table which is then referred to in text covering a number of paragraphs or pages. Small tables are easier to position close to their verbal summary, and are easy to include in the main report. They are also more digestible than large tables.

The standard layout, described in Chapter 5 and summarised in points **a** to **d** below, should be used for the edgings of the table.

2.9 Guidelines for reference tables

The criterion for a good reference table is that it should be easy to use: it should be clearly and legibly laid out and the data should be precisely defined. The first four guidelines on layout apply to demonstration tables as well as to reference tables.

a All the information necessary to interpret the table should appear in the table: this can be achieved by using the standard layout (described in Chapter 5) to provide clear information on:

- the kind of objects enumerated
- the factors by which they are tabulated
- the units of measurement
- the geographical coverage
- the time period covered
- the source of data.

b Use horizontal lines, blank space and shading to guide the eye. Columns and rows should be regularly spaced and as close together as is consistent with clarity. Horizontal lines may be used to separate column headings and a total row from the body of the table. Average rows and/or columns and total rows and columns should be separated from the rest of the table by extra blank space.

Long columns should be divided into blocks of about five entries by an artificial break (a blank row).

Columns in the body of a table should not be separated by vertical lines.

c Tables should be printed upright on the page: if a large reference table extends over two pages, consider repeating the row headings at the right hand side.

d Chapter 3 (pages 12 to 29) gives concise, detailed guidelines on:

- choice of typeface and use of capital letters
- table titles and row and column headings
- numbering of tables
- punctuation marks for large numbers
- layout and use of footnotes.

e Choice of categories to show separately: where possible use a standard set (for example, the Standard Industrial Classification): if in doubt consult the users of the tables. Be consistent from one year to the next and throughout sets of tables on related topics.

f Put in columns the items which will be compared with each other most frequently: it is easier to scan up and down a column than along a row.

g Indication of probable size or error: consider the needs of the likely users of the tables. For general use the figures may be rounded so that they are accurate to the last non-zero digit shown. For more specialised use it is better to give the probable size of errors explicitly either by:

- attaching quality labels to each row or column (for example, label A might indicate ranges of error of less than one per cent, label B might indicate possible errors of between one per cent and ten per cent, and so on);

- or giving the standard error associated with each entry separately (though this can lead to a congested table).

h Include full and clear footnotes, using covering notes conservatively (they may not be read: footnotes are more obvious).

i Include a complete set of definitions of the terms used in the table and also a key to abbreviations and special symbols.

j Make sure that a clear account of the methods used to collect the data and compile the table is readily available for technical readers.

k Commentary on reference tables is not normally necessary. If circumstances call for such a commentary it is important to ensure that it is impartial and that the reader will have no difficulty in linking all points made in the commentary with the appropriate source table.

2.10 When to use charts. Demonstration only, never for reference

Charts, graphs, maps and diagrams seldom have a role to play in presenting data for *reference* purposes. Reading numbers from a chart tends to be slow and imprecise. However, a carefully designed chart may be more effective than a table when used for *demonstration*.

Charts have a number of advantages and disadvantages: you should consider using a chart rather than a table when the pluses outweigh the minuses in the following table.

Table 2.1 Charts: plus points and minus points

+	−
Good for communicating non-specific quantitative comparisons	Poor for communicating specific amounts
Attractive to look at: make a report look more interesting	Can be time-consuming and expensive to design and execute well
Likely to appeal more to a general audience than a table of figures	May not be properly assimilated: people need to be trained to interpret any but the simplest sort of graph
General time trends can be shown and compared more effectively using line graphs than using tables	May not be possible to show related trends on the same graph if the orders of magnitude are different
Communicate one or two messages well	Too many messages make a bad chart

It is comparatively easy to produce statistical charts using modern computer and word-processing packages. However, use of the default options seldom leads to an effective end product and many offer embellishments which are distracting rather than helpful to interpretation. Producing a good chart usually requires time and effort.

2.11 What sort of chart?

A bewildering variety of statistical charts and diagrams have been devised, some requiring considerable skill to interpret correctly. Since it is unrealistic to expect a general reader to pore over an unfamiliar chart in order to decode its conventions, only simple, straightforward charts should be used when writing for non-specialists.

The main strength of charts is their ability to depict such comparisons as changes in the composition of a total, changes over time in one or more measurements and changes in the ranked order of a number of measurements. The following table lists the most frequently encountered comparisons and the charts which may be used to illustrate each.

Table 2.2 Which chart for which comparison?

Type of comparison	Possible charts
Components as parts of a whole	Pie chart. Component bar chart.
Change in composition of a total	Pie charts. Component bar charts (vertical or horizontal bars).
Sizes of related measurements	Grouped bar charts (vertical or horizontal bars).
Frequency with which different measurements occurred	Bar charts. Histograms.
Change over time in one or more related measurements	Line graphs. Column bar charts.
Change in ranked order of a set of measurements	Paired or grouped bar charts.
Correlation between two sets of measurements	Back-to-back bar charts. Scatter diagrams.

General guidelines which apply to all charts are:

a Each chart should include a clear definition of:

- the kind of objects represented
- the units and scale of measurement used
- the geographical coverage
- the time period covered
- the source of data.

b The chart should be easy to read: it should be upright on the page and no bigger than necessary for clarity; lines and sections of the chart should be labelled directly rather than via a separate key and, if shadings or colours are used, they should be easy to distinguish from each other even when photocopied.

c The chart should be easy to interpret: this means that the conventions used in constructing the chart should be self explanatory and not distracting (for example, bars of different widths *and* different heights should not be used): it also means that each chart should be accompanied by a clear verbal summary of its main points.

d It is better to use two or three small, simple graphs to build up a story rather than put too much on a single chart.

e If the use of a specialist graph seems to be justified because of the clarity with which it alone can illustrate a particular point, take time and care to train the reader in how to interpret its conventions: give a simple example of how a familiar measurement would appear on it (for example, regular compound interest on a logarithmic scale).

2.12 The role of words in communicating numbers

Words alone should *never* be used to communicate more than two or three simple numbers. But a verbal summary should accompany *every* table and chart used for demonstration purposes. Since most people find numbers alienating and uninteresting, the prose used in summarising quantitative information should be as lucid and readable as possible.

In writing a verbal summary of a table or chart, include only three or four major patterns displayed by the data: do not include a blow by blow account of every row or column and avoid explanations about how the data are defined or tabulated unless it is essential to point out that an apparently dramatic pattern is attributable to a break in the series. (Such information should, of course, appear clearly as a footnote to the table or chart itself.)

In writing a longer report about a statistical investigation or in producing a summarising commentary on a set of reference tables, aim to make the report as readable as possible. Measures which can help to achieve this are:

- putting the conclusions first
- using short sentences with as few qualifying clauses as possible
- avoiding technical terms as far as possible
- using sub-headings freely
- relating the report to familiar information (by using analogies, or by relating the data to items of general knowledge, for example, "the year of the miners' strike").

Finally, the most important rule in communicating quantitative information is to THINK CLEARLY. If you know exactly what your data say you will have little difficulty in communicating the message effectively.

3 Structure and style of tables

3.1 Introduction

Once you have decided to present a set of data in a table, a number of details must be considered. For example, you must decide what title to give the table, where to show the units of measurement, how much space to leave between rows and columns, what footnotes to include and what typeface to use. These are decisions on how the table is to be *structured* and on its *style* of presentation.

This chapter discusses the structure and style of tables. The guidelines offered apply to both reference and demonstration tables.

3.2 Structure

Just as there are rules to be obeyed if a passage of prose is to communicate a clear and unambiguous message, so there are rules associated with constructing a table of numbers. In the first case, the prose must be grammatically correct; in the second case, the table must be correctly structured.

Fortunately the rules associated with table structure are fewer and simpler than the rules of grammar. They concern the *definition* and *layout* of tables and they can be set out in a few pages.

Most experienced presenters of data acquire some basic knowledge of table structure by following good examples or by trial and error. Although this chapter may be of interest to such presenters, it is really designed to help inexperienced presenters produce clear professional tables *every time*.

Table definition

Every table should contain a clear and complete explanation of the figures presented. The reader should be left in no doubt about:

a   the kind of events, objects, people etc which are enumerated
b   the factors by which the data are categorised
c   the units used: absolute numbers, thousands, rates, percentages etc
d   the geographical coverage
e   the time period
f   the source of data (if the document in which the table appears is not itself a primary source).

A common mistake is to rely on separate text to convey some of this information. This can lead to figures in a table being taken out of context and misinterpreted. The standard arrangement used in most Government Statistical Service publications and papers is to include **a** and **b** in the table title, to state **c**, **d** and **e** immediately above the main body of the table and **f** underneath, like this:

Table 3.1A [Kind of objects, etc]: by [classifying factor or factors]

Geographical and time coverage	*Units*
[Data, in rows and columns]	
	Source

For example:

Table 3.1B Women in employment: by age of youngest dependent child

Great Britain *winter 1992-93*	*Percentages* *not seasonally adjusted*
[Data, in rows and columns]	
	Source: Labour Force Survey

Footnotes should be used to record any changes in definition or changes in the method of data collection which may invalidate comparisons between individual rows or columns within the table.

This arrangement works well in practice and is strongly recommended. It provides a useful discipline and gives the top of the table a neat, professional look. The standard layout need not be used always, but presenters should only depart from it with good reason.

Where a table consists of columns measured in different units, each column must be headed with the appropriate units. See for example Table 3.2 taken from the *BT's Directors' report and Financial Statement 1993* (footnotes omitted).

Table 3.2 Analysis of shareholders

Size of shareholding at 31 March 1993	Number of shareholders	Percentage of total	Ordinary shares of 25p each	
			Number of shares held (millions)	Percentage of total
1 – 399	1,241,871	54.0	282	4.6
400 – 799	675,662	29.4	345	5.6
800 – 1,599	288,686	12.6	299	4.8
1,600 – 9,999	87,270	3.8	207	3.4
10,000 – 99,999	2,382	0.1	77	1.2
100,000 – 999,999	1,341	0.1	436	7.0
1,000,000 – 999,999,999 *(a) (b)*	484		3,186	51.5
1,000,000,000 and above	1		1,353	21.9
Total	2,297,697*(c)*	100.0	6,185	100.0*(c)*

Standard layout for two-way tables

Many tables consist of data tabulated according to two (or more) factors: for example, women in employment can be grouped according to the age of the youngest dependent child and then further subdivided according to whether they are working full or part-time. Data displayed like this are said to be *cross-tabulated*: this means that one margin of the table (say, the row headings) records the age of the youngest dependent child, and the other margin (the column headings) records whether the woman works full or part-time. Alternatively, the table may be referred to as a *two-way table* to indicate that it displays data analysed by two separate factors.
The recommended layout for a two-way table is given in Table 3.3.

Table 3.3 [Kind of objects, etc]: by [factor A and factor B]

Geographical and time coverage *Units*

Factor A	Factor B			Total
	Category B1	Category B2	etc	
Category A1				
Category A2				
etc				
Total				

Source

13

The complete table on women in employment (Table 3.4) will serve as an example.

Table 3.4 Women in employment[1]: by age of youngest dependent child and whether working full or part-time

Great Britain   *Percentages*
winter 1992-3   *not seasonally adjusted*

		% women employed		Total[2] =100%
		Full-time	Part-time	(thousands)
Youngest dependent child aged	0- 4	34	64	1 500
	5-10	34	65	1 300
	11-15	44	54	1 100
	All ages	37	61	3 800
No dependent children		66	32	6 800
All women aged 16-59		55	43	10 600

[1] Employees and self-employed   *Source: Labour Force Survey*
[2] Includes women who did not say whether they worked full or part-time, those on government schemes, unpaid family workers and those who did not fully report their employment status.

As before, there is no need to follow the standard layouts slavishly where there is a well-considered reason for using an alternative structure.

Development of the standard layout for use in more complex tables

The standard layout can be extended to display data analysed according to three or even more factors, but this is seldom advisable. Few people will be able to see the overall pattern in a demonstration table if the data are analysed according to three different factors, and mistakes are likely to occur in using such reference tables.

However, where there are special reasons to justify a three-way table, the standard layout can be adapted in a logical fashion. For example, the table in women in employment could be expanded to include age of women as a third factor (see Table 3.5).

Table 3.5 Women in employment[1]: by age, age of youngest dependent child and whether working full or part-time

Great Britain   *Percentages and thousands*
winter 1992-3   *not seasonally adjusted*

Age of woman	Age of youngest dependent child	% women employed		Total[2] = 100% (thousands)
		Full-time	Part-time	
16-24	0- 4	38	58	160
	0-15[3]	40	56	170
	none	70	25	1 800
	All ages	**67**	**28**	**1 970**
25-39	0- 4	33	65	1 200
	5-10	33	66	900
	11-15	45	53	300
	0-15	35	64	2 500
	none	89	10	1 600
	All ages	**56**	**43**	**4 100**
40-49	0- 4	41	55	100
	5-10	35	63	400
	11-15	43	55	700
	0-15	40	58	1 100
	none	58	41	1 700
	All ages	**51**	**48**	**2 800**
50-59	5-10	..	..	10
	11-15	46	54	60
	0-15[3]	47	51	80
	none	47	51	1 660
	All ages	**47**	**51**	**1 740**
Total 16-59	0- 4	34	64	1 500
	5-10	34	65	1 300
	11-15	44	54	1 100
	0-15	37	61	3 800
	none	66	32	6 800
	All ages	**55**	**43**	**10 600**

[1] Employees and self-employed   *Source: Labour Force Survey*
[2] Includes women who did not say whether they worked full or part-time, those on government schemes, unpaid family workers and those who did not fully report their employment status.
[3] Includes a very few women with children in age groups not shown here
.. The number of women sampled was too small for a reliable estimate.

When some figures in a table represent one kind of entity and others represent different kinds, it is good practice to divide the table into separate, clearly labelled parts, each part containing only one kind of object. This can be done using the same

type of structure as in Table 3.5. For example, Table 3.6 from *Social Trends 24* shows the amount of television watched per week in the first section of the table and the percentage of people watching in the second section.

The sections are separated from each other by blank space and each section is clearly labelled to identify the entity being tabulated. The differences between social classes and the very small changes over time (1986-1992) would be much easier to see in Table 3.6 if the time spent watching television had been rounded to the nearest hour. So this table does not merit a ☑

The rule for dividing the table into parts applies just as strongly when the two entities concerned are numbers and percentages or numbers and rates. For example, Table 3.7 clearly suffers from the decision to show numbers and percentages on alternate lines. In Table 3.8 we have presented the same data in a two-part table, and re-ordered and rounded the data. Patterns can be distinguished much more clearly in Table 3.8 than in the original Table 3.7.

Table 3.6 Television viewing: by social class

United Kingdom	Hours and minutes and percentages		
	1986	1991	1992
Social class[1]			
(hours:mins per week)			
AB	19:50	18:51	19:56
C1	23:05	23:56	25:08
C2	26:00	26:57	27:30
DE	33:35	31:56	31:54
All persons	26:32	26:04	26:44
Reach[2]			
(percentages)			
Daily	78	79	82
Weekly	94	94	95

1 See Appendix, Part 5: Social class.
2 Percentage of the United Kingdom population aged 4 and over who viewed TV for at least three consecutive minutes.

Source: Broadcasters' Audience Research Board; British Broadcasting Corporation; AGB Limited; RSMB Limited

Table 3.7 Economic activity and ILO unemployment rates: by age, spring 1992

Great Britain
per cent; numbers in 1000s

Age	Economically active as per cent of population			In employment as per cent of population			ILO unemployed as per cent of labour force		
	All	Men	Women	All	Men	Women	All	Men	Women
All persons 16+	63	74	53	57	65	49	10	11	7
	27,713	15,676	12,037	25,064	13,890	11,174	2,649	1,785	863
All persons of working age[a]	79	86	71	71	77	65	10	12	7
	26,887	15,369	11,518	24,270	13,598	10,671	2,617	1,770	847
50-54	79	89	70	73	80	66	8	10	5
	2,368	1,323	1,044	2,183	1,194	990	184	130	54
55-59	66	78	55	61	70	52	8	11	4
	1,876	1,093	784	1,719	971	748	157	122	35
60-64	38	53	23	35	47	23	8	10	b
	1,040	705	335	961	633	328	79	72	b
65+	6	9	4	5	8	3	5	5	b
	491	307	183	467	292	175	24	15	b
50 to state pension age	69	74	62	63	66	59	8	10	5
	4,949	3,121	1,828	4,535	2,797	1,738	414	324	90
Over state pension age	8	9	8	8	8	8	4	5	3
	826	307	519	794	292	502	31	15	16

Source: Spring 1992 Labour Force Survey - Great Britain

a Women 16-59; men 16-64.
b Fewer than 10,000; estimate not shown.

Table 3.8 Economic activity of older people and ILO unemployment rates: by age and sex

Great Britain, Spring 1992

Age	In employment			ILO unemployed			Total economically active		
	Men	Women	Total	Men	Women	Total	Men	Women	Total
Number of adults (millions)									
50-54	1.2	1.0	2.2	0.1	0.1	0.2	1.3	1.0	2.3
55-59	1.0	0.7	1.7	0.1	-	0.2	1.1	0.8	1.9
60-64	0.6	0.3	1.0	0.1	-	0.1	0.7	0.3	1.0
65+	0.3	0.2	0.5	-	-	-	0.3	0.2	0.5
50 to state pension age	2.8	1.7	4.5	0.3	0.1	0.4	3.1	1.8	4.9
Working age[1]	13.6	10.7	24.3	1.8	0.8	2.6	15.4	11.5	26.9
Over state pension age	0.3	0.5	0.8	-	-	-	0.3	0.5	0.8
Total adults (aged 16+)	13.9	11.2	25.1	1.8	0.8	2.6	15.7	12.0	27.7
Percentage of adults in age groups	as % population			as % labour force			as % population		
50-54	80	66	73	10	5	8	89	70	79
55-59	70	52	61	11	4	8	78	55	66
60-64	47	23	35	10	..	8	53	23	38
65+	8	3	5	5	..	5	9	4	6
50 to state pension age	66	59	63	10	5	8	74	62	69
Working age[1]	77	65	71	12	7	10	86	71	79
Over state pension age	8	8	8	5	3	4	9	8	8
Total adults (aged 16+)	65	49	57	11	7	10	74	53	63

[1] Working age is 16-59 for women and 16-64 for men
- Less than half the final digit shown
.. Not available (that is can't be calculated reliably for under 10,000 people)

Source: Labour Force Survey

Demonstration tables should seldom contain different entities. Where it is desirable to tabulate two sets of related data, or to show both percentages and original numbers, this will often be done more effectively by using separate tables than by expanding a single table.

Sub-totals

Sub-totals pose an awkward presentational problem. Take, as a simple example, a table showing number of journeys to work, analysed by method of transport and length of journey. There are four main categories of method of transport - private vehicle, public transport, walking and other. But private vehicle contains the sub-categories car, motor-cycle and bicycle And public transport contains the sub-categories bus and train. How do you construct an easy to use table showing all the figures for categories and sub-categories? Clearly it would not be satisfactory to simply list all the figures as over:

Method of transport	Length of journey	
	Under 3 miles	3 to 6 miles
Private vehicle	146	etc
Car	83	
Motor cycle	37	
Bicycle	26	
Public transport	122	
Bus	94	
Train	28	
Walking	66	
Other	11	
Total	345	

A presentational device is needed to show that the figures 146, 122, 66, 11 and 345 have a different status from the others. The best approach is to enter these figures in bold print and use appropriate indentation and row spacing as in the table below.

Method of transport	Length of journey	
	Under 3 miles	3 to 6 miles
Private vehicle		
Car	83	etc
Motor cycle	37	
Bicycle	26	
Total	**146**	
Public transport		
Bus	94	
Train	28	
Total	**122**	
Walking	**66**	
Other	**11**	
Total	**345**	

It is usually better to show totals below their components because this is where we are trained to look for the total of an addition sum. However in reference tables or where space is at a premium, the sub-totals may be shown above the components on the same line as the sub-heading.

A light background shading may be used instead to highlight totals and sub-totals.

Summary of rules

1 State the entities tabulated and the factors by which they have been grouped in the table title.

2 Use the margins of the table to include clear information on:

   • units of measurement
   • geographical and time coverage
   • source of data.

3 If different entities are included in the same table, separate the table into two or more parts, labelling each part clearly to show what entities are tabulated in it and what are the units of measurement. Consider dividing tables which contain two or more entities into separate tables.

4 Indent row headings to indicate entries which are components of totals and sub-totals.

3.3 Applying the rules on structure

Simple as they are, the rules suggested for the construction of statistical tables represent the essence of good professional practice. The reason for most people producing tables of a standard far below that expected of a good professional statistician is that they do not know these simple rules. Consider, for example, Table 3.9, which is taken from an official report and compiled by someone who clearly knew his or her material well, but failed to present it effectively.

Table 3.9 Extract from the Public Service Commissioner's annual report 1991-2

There was a substantial decline in vacancies filled over the financial year. Two hundred and two SES vacancies were filled during 1991-92 as a result of advertised vacancies.

This was a decrease on the previous year's total of 367. Table 2 lists SES vacancies filled as a result of advertising during the period 1988-89 to 1991-92.

Table 2: Total advertised Senior Executive Service vacancies filled 1988-89 1991-92

Method filled	Male/ Female	88-89	89-90	90-91	91-92
Promotions between agencies	M	37	37	19	17
	F	10	12	13	3
Promotions within agencies	M	317	203	222	104
	F	52	29	44	21
Total promotions	**M**	**3 5 4**	**2 4 0**	**2 4 1**	**1 2 1**
	F	**6 2**	**4 1**	**5 7**	**2 4**
Transfers between agencies	M	10	5	6	4
	F	1	2	5	5
Transfers within agencies	M	28	23	20	13
	F	7	6	4	4
Total transfers	**M**	**3 8**	**2 8**	**2 6**	**1 7**
	F	**8**	**8**	**9**	**9**
Appointments	M	17	24	20	11
	F	0	4	6	8
Term appointments	M	4	3	7	9
	F	3	1	1	3
Total appointments	**M**	**2 1**	**2 7**	**2 7**	**2 0**
	F	**3**	**5**	**7**	**1 1**
Totals	M	413	295	294	158
	F	73	54	73	44
Total vacancies filled		**4 8 6**	**3 4 9**	**3 6 7**	**2 0 2**

These figures do not include SES vacancies which were filled without advertising, such as transfers within agencies where advertising of a vacancy did not occur.

The layout of this table is so bad that no-one is going to attempt to read it unless they absolutely have to and then it will take them many minutes to get a proper grasp of what these figures represent. But suppose the author had applied the simple rules summarised in section 3.3. He or she would have followed a chain of thought similar to this:

Q1 What am I enumerating?
A1 Senior Executive Service vacancies advertised and filled.

Q2 By what factors are the data analysed?
A2 There are three factors: time, sex and method of filling the vacancy (although male plus female totals are not given except in the grand total).

Q3 Following on from **A2**: is a full three way analysis really required or would two two-way tables suffice: by sex over time and by method over time? If the full analysis is required for reference, might it be helpful to summarise the totals in a simple demonstration table and give the full analysis separately? Would male plus female totals be useful for each method of filling the vacancy?
A3 Two tables is a good idea and if the total of males plus females is in the first table, the second table need not repeat these totals - saving space is helpful on the small page.

Q4 Is this a reference table?
A4 Yes, so it would be wrong to round the data.

Q5 How can I help the reader understand the structure of 'methods filled'?
A5 Show explicitly how totals are calculated.

Q6 Would percentages, ratios or index numbers be helpful or are numbers sufficient?
A6 Numbers are sufficient.

Q7 What graphic design features are available?
A7 Bold type or shading. The typeface and print size determined by the house style for text. Not easy to produce a chart. No colour.

Had the compiler of Table 3.9 taken this systematic approach, and known the standard layout, he or she would probably have produced instead something resembling the clearer, more professional Tables 3.10 and 3.11. The important point demonstrated by this example is that thoroughly bad tables should never be produced. To present the same data in a form which shows proper respect and consideration for the reader only requires thought and the application of a few simple rules.

Table 3.10 Senior Executive Service vacancies advertised and filled[1]: by sex

1988-89 to 1991-92 *number*

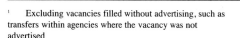

	Male	Female	Total
1988-89	413	73	486
1989-90	295	54	349
1990-91	294	73	367
1991-92	158	44	202

[1] Excluding vacancies filled without advertising, such as transfers within agencies where the vacancy was not advertised

There is, of course, no one 'right table' for any given set of data. There may be several different structures that will present the data clearly and accurately, but using the rules discussed in this chapter will ensure that such a table is produced. However, just as it is worth drafting and revising a passage of writing, it is worth trying several different designs of table in order to decide which one displays the data most effectively.

Table 3.11 Senior Executive Service vacancies advertised and filled[1]:
by sex and method filled

1988-89 to 1991-92   *number*

	Promotions			Transfers			Appointments			Total
	Within agencies	Between agencies	Total (cols 1+2)	Within agencies	Between agencies	Total (cols 4+5)	Permanent	Term	Total (cols 7+8)	(cols 3+6 +9)
Male										
1988-89	317	37	354	28	10	38	17	4	21	413
1989-90	203	37	240	23	5	28	24	3	27	295
1990-91	222	19	241	20	6	26	20	7	27	294
1991-92	104	17	121	13	4	17	11	9	20	158
Female										
1988-89	52	10	62	7	1	8	0	3	3	73
1989-90	29	12	41	6	2	8	4	1	5	54
1990-91	44	13	57	4	5	9	6	1	7	73
1991-92	21	3	24	4	5	9	8	3	11	44

[1] Excluding vacancies filled without advertising, such as vacancies within agencies where the vacancy was not advertised

3.4 Style

Style plays the same role in statistical tables as in language. The first requirement for a table is that its basic structure (grammar) should be correct; the second is that its style should be effective for the task in hand. A particular type of style may be good in one application but bad in another, depending on the information in the table and the customers who are to use it. The test of good style is its success in getting information across correctly, economically and, where necessary, powerfully. If a table works well, its style is, by definition, good. Nevertheless rules of style do exist. They do not have the self-evident correctness of the rules of structure and the reasons for proposing them vary: some have been checked under controlled experimental conditions while others owe their importance to the fact that they have been found, in practice, to work well most of the time. They are best thought of as guidelines to follow unless there is a sound reason for not doing so.

Before discussing these rules, let us clear the ground by considering in more detail the two distinctive roles that tables can play in statistical presentation.

Reference tables and demonstration tables

In Chapter 1 we explained that statistical tables can serve two quite different purposes:

a provision of data for reference (essentially a storage function)
b demonstration of particular facts or situations (essentially a means of influencing decisions).

The main aim in **a** is to provide figures which the user will find it easy to look up and understand; in **b** it is to get particular messages across to the reader.

The style of a table will depend strongly on whether it is for reference or for demonstration purposes. For example, if a table giving the population of counties in England were intended for reference, the counties would be best listed in alphabetical order within standard region whereas, for demonstration purposes they would probably be better shown in some other order, such as population size.

Wherever possible, tables should be dedicated wholly to purpose **a** or wholly to purpose **b** and the style of presentation should be chosen accordingly. Any attempt to serve both purposes in the same table should be resisted: it is most unlikely that the same table will serve both purposes well and the danger of ending up with a table which serves neither properly should be borne in mind. If some such attempt *must* be made, decide which is the *primary* role of the table and design the table to carry out that role effectively. Then consider how the table would have to be changed in order to fulfil the secondary role effectively, for example, by rounding all the numbers in a reference table for demonstration purposes. At that point a compromise may have to be reached.

Factors affecting 'style' are:

*   layout: spacing of rows and columns and use of ruled lines, frames and shading
*   choice of typeface and use of bold and capital letters
*   numbering of tables
*   punctuation marks for large numbers
*   footnotes
*   rounding
*   colour and shading

Layout

A very common mistake is to allow the spacing of rows and columns to be determined chiefly by the length of row captions and column headings. This usually results in a table that is untidy and difficult to read. It is nearly always possible to shorten or rearrange the captions and headings so that they do not interfere with spacing.

Another common mistake is to have columns too widely separated. This usually arises because typists and designers, unless otherwise instructed, are likely to space the columns out to cover the full width of the page. (See Table 5.7 and almost all company reports!) Irrespective of the number of columns, spaces between them need only be just big enough to prevent confusion of figures in one column with those in another and to allow the eye and mind to identify figures in individual columns as distinct groups. Somewhere between five and eight character units of blank space between figures in neighbouring columns is generally about right. Wider spacing should only be used to separate distinct sets of columns, for example to separate the total column from its component columns as in Table 3.11.

Much of what has been said for columns also applies to rows. In general, rows of figures should be spaced no further apart than normal rows of text. Wider spacing should only be used to separate distinct sets of rows or to create artificial 'breaks' in large blocks of consecutive rows. Those breaks should be inserted after every fourth or fifth row. This simple device enables the eye to identify a row much more quickly and reliably than it could if all the rows were equally spaced.

A typical spacing pattern for the numbers in a medium-sized two-way table would therefore be something like this:

						(Total)
	x	x	x	x	x	x
	x	x	x	x	x	x
	x	x	x	x	x	x
	x	x	x	x	x	x
(Artificial break)						
	x	x	x	x	x	x
	x	x	x	x	x	x
	x	x	x	x	x	x
	x	x	x	x	x	x
(Total)	x	x	x	x	x	x

Table 3.12 The European climate

Changes in certain climatological data, by country							
	Number of stations	Rainfall		Average temperature over the year		Water balance	
		(mm H$_2$O)		(°C)	(°C)	(mm H$_2$O)	
		Min.	Max.	Min.	Max.	Min.	Max.
B	2	824	1 191	7.2	9.7	294	647
DK	4	551	874	7.5	8.1	− 35	243
D	19	560	979	7.8	10.6	− 155	373
GR	8	422	1 044	14.1	18.2	:	:
E	25	218	1 702	11.0	18.5	− 635	− 122
F	37	493	1 236	9.5	14.9	− 280	554
IRL	6	781	1 502	9.0	10.6	160	861
I	23	424	1 142	11.5	18.4	− 418	542
L	1	856	:	8.4	:	271	:
NL	5	562	777	8.5	9.7	− 292	199
P	7	527	1 064	13.2	17.2	− 742	217
UK	21	496	967	8.4	10.9	− 93	442

Such a pattern is not possible if some of the row captions take up more than one line: wherever possible, row captions should be made short enough to fit on one line.

An alternative method of ensuring that it is easy to read across figures in the same row is to shade alternate rows as in Table 3.12.

On the question of ruled lines (called 'rules' in printing terms), the basic principle is to use no more than necessary. They create extra 'clutter' in a table and detract from its visual appearance. If proper care is taken with table layout and row and column spacing, three horizontal lines will usually be sufficient: occasionally up to five may be used as follows:

(Table heading)

	(Coverage)				(Units)
Line 1 -					
	Factor A	Factor B			Total
Line 2 -					
		B1	B2	B3	B4
Line 3 -					
	A1	x	x	x	x	x
	A2	x	x	x	x	x
	A3	x	x	x	x	x
	A4	x	x	x	x	x
Line 4 -					
	Total	x	x	x	x	x
Line 5 -					

(Source)

Each line serves a purpose; further lines would be superfluous. Given careful control over row spacing and use of bold print for totals, line 4 and sometimes line 3 can be omitted without loss of clarity (as, for example in Table 3.4). Some people prefer lines 3 and 4 to be solid, rather than broken; this is a matter of personal taste. Sometimes the general column headings 'Factor A' and 'Factor B' can be omitted without ambiguity and line 2 is not then required. All the example tables in this chapter (except those given as examples of poor layout or style) demonstrate correct use of ruled lines.

Some presenters prefer to avoid using vertical lines ('rules') altogether, while others favour using them either to box in the table, or to separate a 'Totals' column from the body of the table, or both. This is another matter of personal taste. However, vertical lines should never be drawn between columns in the body of a table: this merely prevents the eye moving smoothly across the rows.

Tables should be printed upright on the page and not sideways: it is irritating to have to twist a book or report in order to read a table (or chart). Occasionally there may be a sufficiently good reason to break this rule for reference tables, but it should never be broken when designing a demonstration table.

Choice of typeface and use of bold and capital letters

The objective here is to adopt a style which is easy to read, pleasing to look at and consistent throughout. The choice of typeface is a personal one but plain kinds are generally best. Within the same publication, all tables (and charts) should conform to the same pattern used. In practice, the options available will depend on whether the table is to be professionally typeset, produced at the workplace using a computer with a printer attached, or typed traditionally. However, with care and attention to detail, satisfactory tables can be produced with any configuration of hardware and software.

The following guidelines are offered as one possible recipe for achieving a consistent and effective style.

All tables must have a clear informative *title* as described in section 3.2, and tables should generally be numbered (see later paragraphs on *Numbering of tables*). Where a table is numbered, its title should be above the table, immediately following the table number but separated from it by a space: where a second line is needed it should begin under the first word of the title, not under the table number.

The title should be in lower case letters using capital letters sparingly. The table title should be produced in bold print and, where possible, in a larger typeface than that used in the main body of the table. Capitals should be used only for the first letters of the initial word and of proper nouns. If the same size and weight of typeface has to be used for the title as for row and column headings, then capital letters should normally be used for the initial letters of all but minor words in the title. (Minor words include words like: of, by, and, the, etc.)

Row and column headings should be in the same typeface as the rest of the table, and bold or italic type should be used only when needed to make distinctions, for example, to indicate major sub-headings and totals.

In row and column headings only the initial word and proper names should begin with a capital letter.

As far as possible column headings should be horizontal and not sideways on.

Where column headings are of varying length and take up varying numbers of lines all the column headings should begin on the same line and those of fewer lines should not be centred in the vertical space.

If a row description runs to two or more lines, the second and subsequent lines(s) should not be inset.

Where a row category is further divided into a number of sub-categories, this should be indicated by in-setting the row headings. See, for example, Table 3.4.

Statistical tables are usually printed in either ten point or twelve point type. The choice is largely a matter of personal preference. Compare Tables 3.13 and Table 3.14. Table 3.13 is in 10 point type, which is used throughout this book and Table 3.14 is in 12 point type. Some experiments have been carried out[1] to establish which size is easier to read accurately and the results have always been marginally in favour of the larger typeface. However no difference has been great enough for the experimenter to be sure that the apparent advantage was not due to random variation between the people taking part in the experiment.

It is important that all the numbers should be legible and that the table should look neat. Some people find that a smaller table often looks neater. Compare Table 3.14 (printed in 12 point type) with Table 3.15 - the same table reduced to 70 per cent of its original size using a photocopier. (This is the reduction necessary to reduce A3 to A4.) The most suitable size for any particular display is best decided on by experimentation.

Footnotes will normally be printed using a smaller type size. It is important to ensure that, whatever size of typeface is chosen for figures, any numbers which appear in footnotes should remain legible.

The use of bold type for totals and sub-totals was discussed in section 3.2. It is sometimes useful to employ italics for percentages and rates when these appear in the same table as actual numbers (although the distinction between plain and italic numbers is not always clear). In Table 3.6 for example, percentages are shown in italics. If, as was recommended in section 3.2, percentages and rates are assembled in separate parts of the table, the need for a different typeface is less pressing.

Numbering of tables

Table numbers may be consecutive throughout a publication or alternative methods may be used. Where convenient, separate sets of tables may be numbered in sections. For example, tables may be numbered first by the chapter number and then by their position within the chapter (as in this book). Clearly the same number must not be used twice in the same publication.

Tables, graphs and figures are traditionally given separate series of numbers. The series of numbers may be reduced to two by having one series for tables and another for figures which will include graphs, maps, charts and pictures.

Some publications, such as *Social Trends*, combine tables and charts in a single series, labelling items in very large type as simply '6.2' and referring to the item as either 'Table 6.2' or 'Chart 6.2' as appropriate.

It may not be necessary to number every demonstration table. If a table of data appears in the middle of its supporting text *and* if it is not referred to at any other point in the report then a table number is not strictly necessary. In practice, however, it is often desirable to refer back to a data display and this can be done easily if every data display is numbered. If you decide not to number a particular table and, later, reverse this decision in order to refer back to it, the process of renumbering subsequent tables and adjusting references to them is time-consuming and prone to error.

Punctuation marks for large numbers

Punctuation marks of some kind are essential in large numbers, to mark the position of digits representing thousands and millions. (1,270,000 should never be entered in table as 1270000.) Although a comma is customarily used in the UK for this purpose, a better method is to omit the comma itself but space the digits as though the comma were there. The resulting small gaps will be quite sufficient to separate the thousands from the hundreds, and the absence of commas will produce a clearer table.

---

[1] TINDER M A *Legibility of mathematical tables*. Journal Applied Psychology, 1960, vol 44, no2, pp83-87

Table 3.13 Composition of consumer credit (10 point type)

United Kingdom   *Percentages and £ billion*

	Credit source (percentages)						Credit outstanding at year end (£ billion)
	Bank loans[1]	Bank credit card lending	Finance houses[2]	Retailers	Insurance companies	Building society loans[3]	
1982	67	13	8	10	2	..	16
1987	63	17	12	6	2	..	36
1991	63	18	10	5	3	1	54
1992	63	19	9	5	3	1	53

[1] Banks and all other institutions authorised to take deposits under the Banking Act 1987.
[2] Finance houses and other credit companies (excluding institutions authorised to take deposits under the Banking Act 1987).
[3] Building Society unsecured loans to individuals or companies (i.e. Class 3 loans as defined in the Building Societies Act 1986).

Source: Central Statistical Office

Table 3.14 Composition of consumer credit (12 point type)

United Kingdom   *Percentages and £ billion*

	Credit source (percentages)						Credit outstanding at year end (£ billion)
	Bank loans[1]	Bank credit card lending	Finance houses[2]	Retailers	Insurance companies	Building society loans[3]	
1982	67	13	8	10	2	..	16
1987	63	17	12	6	2	..	36
1991	63	18	10	5	3	1	54
1992	63	19	9	5	3	1	53

[1] Banks and all other institutions authorised to take deposits under the Banking Act 1987.
[2] Finance houses and other credit companies (excluding institutions authorised to take deposits under the Banking Act 1987).
[3] Building Society unsecured loans to individuals or companies (i.e. Class 3 loans as defined in the Building Societies Act 1986).

Source: Central Statistical Office

Table 3.15 Composition of consumer credit (Table 3.14 reduced to 70 per cent size)

United Kingdom   *Percentages and £ billion*

	Credit source (percentages)						Credit outstanding at year end (£ billion)
	Bank loans[1]	Bank credit card lending	Finance houses[2]	Retailers	Insurance companies	Building society loans[3]	
1982	67	13	8	10	2	..	16
1987	63	17	12	6	2	..	36
1991	63	18	10	5	3	1	54
1992	63	19	9	5	3	1	53

[1] Banks and all other institutions authorised to take deposits under the Banking Act 1987.
[2] Finance houses and other credit companies (excluding institutions authorised to take deposits under the Banking Act 1987).
[3] Building Society unsecured loans to individuals or companies (i.e. Class 3 loans as defined in the Building Societies Act 1986).

Source: Central Statistical Office

Footnotes

Footnotes are a necessary evil. Inches of footnotes after a modest table may produce an unbalanced effect and are likely to be ignored; on the other hand it is essential that all the information necessary for the correct interpretation of the table should be included in the table design. Footnotes will often therefore be necessary to:

- give fuller definition of a brief heading
- explain any changes in definition which have affected the data in the table
- explain any apparent anomalies or inconsistencies in the table.

Fewer footnotes are needed for a demonstration table which has been condensed and rounded to present a broad picture; it does not need to record the full minutiae of the figures. As with rounding, the acid test is whether an incorrect inference might be drawn if the footnote were dropped.

Footnotes should be listed one under the other and placed at the extreme left below the table. The style of type for the footnotes should be the same as for the table but smaller if that is convenient and the size is still above the minimum for readability.

Footnotes should be referred to by superior arabic numbers. Letters in brackets or parentheses may be used as an alternative when needed to avoid confusion. Special symbols, such as asterisks and daggers may also be used. However these symbols are not suitable when more than two or three footnotes are required.

The indicator of the footnote should be above and to the right of the relevant entry, and the type size for the reference symbol should generally be smaller than for figures in the tables.

Footnotes should be referenced across the columns starting from the left-hand side and the top row - which may be the title - and proceeding similarly for each successive row.

The footnote 'Components may not add to totals (or the total) because they have been rounded independently' should be marked against the appropriate headings.

Colour

Tables are traditionally set in black and white with the use of colour, if any, restricted to charts. But as colour becomes cheaper and begins to appear routinely in publications of all kinds, it is also beginning to appear in tables.

The simplest case is where the publication is set, not in traditional black, but in another colour. This choice is a question of taste and 'house style' but bear in mind that light colours may not photocopy well and that a reader may be able to guess the sense of barely legible *text*, but not the content of a table. (See for example Table 3.16.)

Table 3.16 Calls by fire brigades

TABLE 14.ii TOTAL CALLS BY FIRE BRIGADES (ENGLAND AND WALES)

	1987	1988	1989	1990 (revised)	1991
Fire	291,984	296,499	384,379	400,904	356,500[1]
False fire alarm	267,774	285,599	325,777	351,849	367,500[1]
Special service	175,846	139,768	160,090	190,787	169,000[1]
Total	735,604	721,866	870,246	943,540	893,000[1]
Calls per firefighter	14.4	14.2	17.1	18.4	17.6

[1] *Rounded and provisional.*

TABLE 14.iii NUMBER OF FIRE CALLS (ENGLAND AND WALES)

	1987	1988	1989	1990 (revised)	1991
Fires involving property or casualties	142,842	143,599	152,943	154,902	165,500[1]
Chimney fires	33,859	27,787	24,560	20,982	24,600[1]
Small fires, including grass and heath land	115,283	125,113	206,876	225,020	166,400[1]
Total fire calls	291,984	296,499	384,379	400,904	356,500[1]

[1] *Rounded and provisional.*

The second simple use of colour is to provide a shaded box as background for the table as in many tables in this book. The shading must be very light as it is here, otherwise the figures merge into the shading; this is a particular problem when a table is photocopied as the shading tends to become darker in the photocopying process. The *Labour Force Survey Quarterly Bulletin* makes more adventurous use of colour in its reference tables, mainly to guide the reader through the layout of the table, but also to highlight certain figures in red. In Table 3.17 the percentage changes between Winter 1992 and Winter 1993 are shown in black if they are positive, but red if they are negative. This use of colour to highlight key figures is frequently used in oral presentations, and, when used properly, can make both demonstration and reference tables more immediately comprehensible.

As with all graphic design, it is the author's responsibility to ensure that the design enhances the table without overwhelming it as, for example, in Table 3.12.

3.5 Applying the guidelines on style

The use of these guidelines is illustrated in Table 3.18 taken from *Social Focus on Children*.

In this table the title indicates clearly that the table shows children categorised by ethnic group and sex. The data refer to Great Britain in 1991; the entries are in thousands and the source of information was the *Office of Population Censuses and Surveys and the General Register Office (Scotland)*. The footnote explains that those people who described themselves as 'Black British' have been included in 'Black Other'.

Table 3.17 Extract from the Labour Force Survey Quarterly Bulletin

EMPLOYMENT
Table 10: Average actual weekly hours of work (by industry sector) Great Britain

Not seasonally adjusted   SIC (92) (Standard Industrial Classification)

	All Employees & Self-employed[12]	Agriculture and Fishing A-B	Energy and Water C,E	Manufacturing D	Construction F	Distribution, Hotels & Restaurants G-H	Transport I	Banking, Finance & Insurance etc. J-K	Public admin, education & health L-N	Other Services O-Q	Total Services G-Q
ALL											
Winter 1992/3	32.2	41.5	36.2	34.5	35.3	31.0	36.9	32.2	28.5	29.6	30.8
Spring 1993	33.6	44.3	37.4	36.9	38.5	31.9	38.0	33.3	29.1	30.7	31.6
Summer 1993	32.4	44.5	37.0	35.1	38.0	31.6	37.7	32.3	26.9	30.0	30.6
Autumn 1993	33.9	43.5	38.1	37.3	39.5	31.9	38.5	33.6	29.6	30.0	31.9
Winter 1993/4[3]	32.2	41.2	35.1	34.9	35.7	31.1	38.0	32.6	28.4	28.1	30.8
Change											
Win92-Win93	*0.0*	*-0.3*	*-1.1*	*0.4*	*0.4*	*0.1*	*1.0*	*0.4*	*-0.1*	*-1.5*	*0.0*
MEN											
Winter 1992/3	37.6	45.1	37.5	36.9	36.6	38.9	40.0	37.1	35.9	36.2	37.8
Spring 1993	39.5	48.6	39.2	39.6	40.0	40.4	41.0	38.4	36.5	38.0	38.9
Summer 1993	38.3	49.3	38.9	37.6	39.7	39.8	40.5	37.5	34.0	37.2	37.8
Autumn 1993	39.9	48.0	40.2	39.9	41.1	40.4	41.4	39.4	37.0	37.1	39.3
Winter 1993/4[3]	37.7	44.9	36.9	37.3	37.0	38.9	40.2	38.1	35.4	34.9	37.8
Change											
Win92-Win93	*0.1*	*-0.1*	*-0.6*	*0.4*	*0.4*	*0.0*	*0.2*	*1.0*	*-0.5*	*-1.4*	*0.0*
WOMEN											
Winter 1992/3	25.6	27.6	30.4	28.4	24.2	24.1	27.5	26.8	25.0	24.5	25.1
Spring 1993	26.4	29.5	30.4	30.1	25.9	24.5	28.4	27.8	25.6	25.0	25.7
Summer 1993	25.2	27.6	29.3	28.7	23.9	24.3	28.6	26.7	23.6	24.6	24.6
Autumn 1993	26.5	26.7	29.8	30.4	25.7	24.4	28.7	27.4	26.2	24.8	25.9
Winter 1993/4[3]	25.5	28.5	27.8	29.0	23.1	24.0	29.8	26.5	25.1	23.0	25.0
Change											
Win92-Win93	*-0.1*	*0.9*	*-2.6*	*0.5*	*-1.2*	*-0.1*	*2.3*	*-0.3*	*0.1*	*-1.5*	*-0.1*

NOTE: Average hours actually worked in the reference week

Table 3.18 Children: by ethnic group and age, 1991

Great Britain
Thousands

	0-4	5-9	10-15	All aged under 16	As a percentage of total population
Ethnic minority group					
Black Caribbean	38	36	36	109	21.9
Black African	25	19	18	62	29.3
Black Other[1]	36	29	25	90	50.6
Indian	74	82	92	248	29.5
Pakistani	63	67	73	203	42.6
Bangladeshi	25	25	27	77	47.2
Chinese	11	12	14	37	23.3
Other Asian	16	16	16	48	24.4
Other	48	39	35	121	41.7
All ethnic minority groups	335	324	337	996	33.0
White	3,299	3,116	3,613	10,027	19.3
All ethnic groups	3,633	3,440	3,950	11,024	20.1

1 Black Other includes those who answered Black British.

Source: Office of Population Censuses and Surveys; General Register Office (Scotland)

There are two major categories in the table rows, ethnic minority groups and white; these labels are aligned with the left-hand edge of the table. The different ethnic minority categories are shown below the 'ethnic minority' label and are slightly, but clearly, indented. These nine categories are grouped into three natural groups: black, Indian sub-continent and other giving two natural breaks in the data. The total for all ethnic minorities is given below the components, separated by a blank row for clarity.

The table headings are justified to the bottom of the column, which makes it slightly more difficult for the reader to scan the headings.

On closer examination we find that all the entries are in thousands apart from the final column which is in percentages. We would have preferred the columns in thousands to have been labelled directly. The entries in the final, percentage, column should have been rounded to the nearest whole number and a device such as italics, or extra space might have helped to distinguish them from the members in the rest of the table.

The headings and labels are both printed in a slightly darker shade for emphasis. The shade chosen is not so dark that it causes the headings and labels to dominate the table.

There are three ruled lines, which are also darker than the text, but not too obtrusive. In fact two lines would have been sufficient. The line under the column headings need not run the full width of the table and is not strictly necessary.

The columns are regularly spaced (despite different lengths of column headings) and as close together as can be achieved given that the final column heading includes two long words: 'percentage' and 'population'.

The font chosen is sans-serif (apart from our title), which means the characters are plain rather than embellished, and this contributes to the uncluttered appearance of the table. This book in contrast is printed in a serif font. The print size is perfectly legible as is the footnote which is in smaller type.

Commas are used to punctuate numbers running into thousands. As there are few large numbers this does not add significantly to the amount of ink on the paper, but we recommend that the commas be replaced by small spaces in printed tables.

Overall the table is well laid out, uncluttered and attractive in appearance and is clear and easy to interpret.

The last table in this chapter illustrates what can be achieved using a spreadsheet package on a desktop PC. Table 3.19 is clear and uncluttered with lines generated by the spreadsheet package, properly aligned headings and a plain typeface (Arial 10 point). One minor fault is that the footnote indicators are too large and have had to be put in brackets to distinguish them from the row headings. Another common problem has been avoided here by rounding the total to two effective digits: commas, rather than spaces, have to be used to separate thousands and hundreds. Both these faults can be avoided if tables are generated in a standard word processing package, but a far more serious problem would arise. In most word processing packages it is extremely difficult to align numbers correctly within columns as shown Table 3.19.

Finally, while printing and publishing are outside the scope of this book, anyone presenting numbers must be aware that the choice of fonts and colours, the density of ink, the use of frames and background shading and general 'house style' rules can all significantly affect the look of a table or publication. The examples in this book illustrate some of the features available. With desk top publishing, camera-ready copy and modern printing techniques, the author has more control over the appearance of his work than ever before and should ensure that modern technology is used to enhance communication rather than to produce an eye-catching muddle. Where graphic designers are brought in to achieve a professional image it is important to remember that they may not be aware of the principles of designing good tables and charts. It is the author's responsibility to ensure that the graphic design enhances the effective presentation of the patterns in the data.

Table 3.19 Motor vehicles currently licensed (spreadsheet version)[1]

Great Britain							thousands
	Type of vehicle						All vehicles (2)
	Private cars	Other private & light goods vehicles	Motor cycles etc	Goods vehicles	Crown and exempt vehicles (3)	Other vehicles	
1953	2,000	500	900	450	90	420	5,000
1963	6,000	1,100	1,500	540	120	580	10,000
1973	12,000	1,600	900	540	140	560	15,000
1983	16,000	1,700	1,300	490	620	580	20,000
1993	(4) 20,000	2,200	700	430	980	480	25,000
Average	11,000	1,400	1,100	490	..	530	15,000

(1) Taxation classes changed in 1990. Figures for earlier   Source: Department of Transport
years have been estimated on the new basis.
All classes except motor cycles were affected, particularly the private cars and other private and light goods vehicles classes.
(2) Components may not add to the total because they have been rounded independently.
(3) Coverage increased in the 1980s to include some vehicles previously exempt from tax . There were 170,000 in 1981. A meaningful average cannot be calculated because of the change.
(4) Method of estimation changed in 1992. Estimates for 1992 were 1% lower on the new basis than on the old basis.

4 Demonstration Tables

4.1 Which numbers to present

In demonstration tables the prime function of the table is to *communicate a message*. This means that you, the writer, must first decide what message to present by careful analysis of the data, and then design a table to illustrate that message as effectively and memorably as possible.

Most data can be presented in a number of different ways: for example, each number can be expressed as a percentage of the appropriate row or column total; data recorded over time can be expressed as a series of index numbers based on a specific year whose value is taken as 100; the ratio or difference between two rows or columns may be calculated; the incidence of accidents or disease may be quoted as rates per 1,000 at risk; sums of money can be recorded either in current terms or can be measured at constant prices by dividing each number by an appropriate factor; and of course, the original data can be displayed, usually rounded to two or three figures.

The choice of which numbers to use in a demonstration table depends on the message being illustrated. In practice the process of analysing the original data is likely to involve calculating a number of derived statistics (differences, ratios, percentages and averages) and, by the time the main patterns in the data have been identified, the most appropriate way of highlighting these patterns is likely to be clear to the analyst.

At the exploratory stage some patterns in the data may be revealed by calculating fairly complicated derived statistics (for example, ratios of successive differences or percentages of percentages) but in presenting the data it is important to avoid using over-elaborate derived statistics.

Few readers will have any difficulty in interpreting a table of percentages (so long as it is clear which total the percentages are based on), or a table of numbers quoted as indices based on a specific year (for example, 1990 = 100) but may will quail before a column headed 'ratio between male and female upper quartiles'.

Section 4.2 lists a number of derived statistics andgives examples of how they can be used effectively. The list is not exhaustive and other derived measures may be found which illustrate particular messages effectively. If in doubt about the effectiveness of a possible table of derived statistics, ask yourself 'will my mother understand this table?'.

4.2 Some derived statistics

Percentages

Percentages are useful when changes in the composition of a total are of particular interest. For example, Table 4.1 shows that males and females tend to take part-time jobs for different reasons and that the pattern is also different for married and non-married females: 88 per cent of married female part-timers did not want a full-time job compared with 45 per cent of non-married female and 36 per cent of male part-timers. For non-married females nearly as twice as many had taken a part-time job because they were a student or still at school rather than because they could not find a full-time job. (33 and 18 per cent respectively). For men, the two reasons were given with the same frequency (both 29 per cent).

Table 4.1   Reasons for taking a part-time[1] job: by sex and marital status, Spring 1993

United Kingdom		Percentages and thousands		
		Females		
	Males	Married	Non-married	All females
Student/still at school	29.4	0.6	33.4	6.9
Ill or disabled	3.3	1.0	1.3	1.1
Could not find a full-time job	29.0	8.4	18.3	10.3
Did not want a full-time job	36.2	88.0	45.4	79.9
Part-time workers[2] (=100%) (thousands)	886	4,078	967	5,045

1 Part-time is based on respondent's self assessment.
2 Includes those who did not state the reason for taking a part-time job.
Source: Employment Department

When a demonstration table shows percentages, the totals on which the percentages are based should always be quoted if they are not given elsewhere. There are two reasons for this: first, quoting the totals allows an interested reader to explore the table more thoroughly by working out the numbers in each category (for example, the number of females working part-time because they could not find a full-time job is actually higher than the number of males - 10 per cent of 5,045 is about 505 compared with nearly 30 per cent of 886 which is about 265); and secondly, it provides necessary background information against which the significance of changes in percentages can be assessed (a change from 30 to 33 per cent on a total of 100 represents only three individuals: the same change in percentages based on a total of 10,000 represents 300 individuals and is clearly a more significant change). Naturally, if the original numbers are of fundamental importance in their own right, they should be displayed as well as the percentages. Tables may contain percentages of either column totals, as in Table 4.1, or row totals, as in Table 4.2. When both the numbers and percentages are displayed, it is generally best to put them in separate tables, or separate parts of the same table.

Decimal points are rarely needed in demonstration tables showing percentages. Tables 4.1 and 4.2 would both be easier to use without the decimal points and this is why neither has been given a ☑ . Authors often recognise this implicitly by rounding the numbers in the verbal summary as is the case with Tables 4.1 and 4.2. This makes the text easier to read, but it can take the reader a few moments to relate 15 per cent in the text to 14.5 per cent in the table. Better to round the table.

Demonstration tables should always be accompanied by a verbal summary highlighting the main patterns. The verbal summaries which accompanied Table 4.1 and Table 4.2 are given below.

Verbal summary to Table 4.1

Most people work part-time out of choice. However, in 1993 nearly 15 per cent accepted part-time work because they could not find a full-time job, compared with just over ten per cent a year earlier. Amongst men and non-married women around three in ten took a part-time job because they were a student or still at school.

Verbal summary to Table 4.2

The income of the household affects its pattern of expenditure. Households with less than £100 of disposable income per week spent nearly a quarter of their expenditure on food and a further quarter on housing, fuel, light and power. Those with over £400 of weekly disposable income, on the other hand, spent only 15 per cent of their expenditure on food and only a further 20 per cent on housing, fuel, light and power. People earning between £200 and £400 spent proportionately more than twice the amount on motoring and fares than people below £100.

Table 4.2 Household expenditure: by level of disposable income, 1992

United Kingdom										Percentages and £ per week
	Percentage of expenditure									Average expend- iture (= 100%) (£ per week)
	Housing	Fuel, light and power	Food	Alcohol and tobacco	Clothing and footwear	Household goods and services	Motoring and fares	Leisure goods and services	Other goods and services	
Normal weekly disposable income										
Under £100	16.0	10.7	23.6	6.8	5.8	14.9	7.7	10.4	4.1	95.21
£100 -£200	16.7	6.9	21.4	7.2	5.6	12.8	12.8	12.2	4.3	173.18
£200 - £400	18.0	4.6	18.1	6.4	6.2	12.7	16.0	13.5	4.5	287.40
Over £400	17.3	3.2	14.5	5.2	6.1	13.0	18.0	18.4	4.3	509.91
All households	17.4	4.8	17.5	6.0	6.0	13.0	15.8	15.0	4.4	271.83

Source: Central Statistical Office

The verbal summary to Table 4.1 is simple and informative, but difficult to relate to the table. The first figure quoted (15 per cent) is not in the table. The reader may take some time to realise that it combines the 29.0 per cent of males and 10.3 per cent of all females who could not find a full-time job. It would also have been reassuring after this, if the author had expanded in 'around three in ten' by adding in brackets (29 and 33 per cent respectively).

The verbal summary to Table 4.2 is workmanlike, interesting and uses rounded numbers. Most of the commentary compares the top and bottom income groups implying that the middle income groups are in between, which is generally true, and no attempt has been made to comment on every column or row.

Indices

When the reader is invited to compare the growth of two or more measurements over a number of years, differences in growth patterns may be hard to detect if each series has a different starting value. For example, Table 4.3 from the *Digest of Environmental Protection and Water Statistics* shows energy consumption over time by six final user groups of different sizes. The table is not particularly easy to use as it stands either as a reference or demonstration table.

It is clear that total energy consumption increased between 1986 and 1992, but the relative changes for different users are unclear. These would become clearer if the table were transposed and rounded and the numbers were lined up correctly (in Table 4.3 the revision indicator (r) has been placed in the units column). However it would still be difficult to compare changes in the transport and public administration categories because the two are very different in size. In Table 4.4 consumption in each year has been expressed as a percentage of the 1986 value. Thus the 1991 volume of energy consumed by final users in industry is given as 94 (that is 15,266/16,208 x 100) and the table is labelled to show that the entries are indices based on 1986=100. The revision indicators have been dropped.

From Table 4.4 it is immediately clear that total energy consumption increased by 5 per cent between 1986 and 1991, with slight falls in consumption by industry, public administration and agriculture and marked increases in consumption by transport and 'miscellaneous' users. We can also see clearly the variation in trends over time and changes in the latest year; the sharp increase in domestic consumption and consumption by 'miscellaneous' users between 1990 and 1991 stand out particularly. Table 4.4 would be an effective demonstration table.

Table 4.3 Energy consumption by final users[1]

United Kingdom

Million therms

	1986	1987	1988	1989	1990	1991
Industry	16,208	16,666r	16,143r	15,759r	15,390r	15,266
Transport	16,258	16,940	18,002	18,834r	19,306	19,044
Domestic	17,348	17,253	16,737	16,073	16,191	17,876
Public administration	3,544	3,405	3,285	3,049	3,045r	3,246
Agriculture	565	544	536	502	504	514
Miscellaneous	3,898	3,921	4,093	4,114	4,177	4,504
Total supplied to final consumers	57,820	58,232	58,796r	58,330r	58,762	60,451

[1] Heat supplied basis

Source: *Digest of UK Energy Statistics, 1992*

Table 4.4 Changes in energy consumption: by final users[1]

United Kingdom *Index numbers: 1986=100*

	Industry	Transport	Domestic	Public admin.	Agric-ulture	Misc.	Total
(bn therms	16.2	16.3	17.3	3.5	0.6	3.9	57.8)
1986	100	100	100	100	100	100	100
1987	103	104	99	96	96	101	101
1988	100	111	96	93	95	105	102
1989	97	116	93	86	89	106	101
1990	95	119	93	86	89	107	102
1991	94	117	103	92	91	116	105

1 Heat supplied basis

Source: *Digest of UK Energy Statistics 1992*

Table 4.5 Shops, number of employees and turnover: by type of retail organisation

Great Britain, 1982-1991

	Shops[1] (thousands)			Turnover[2] (£ billion)			Employment[3] (thousands)		
	Single outlet	Small[4] multiple	Large multiple	Single outlet	Small[4] multiple	Large multiple	Single outlet	Small[4] multiple	Large multiple
1982	220	73	63	22	9	39	850	340	1 070
1987	213	69	63	30	12	63	790	310	1 230
1991	206	66	71	36	16	88	760	310	1 300

1 and other retail outlets eg market stalls and shops within shops
2 including VAT
3 excluding owners and unpaid family members
4 2-9 retail outlets

Source: Business Monitor SDA25
Central Statistical Office

Ratios

Table 4.5 gives the number of shops, their turnover (total sales) and the number of people employed in three different types of retail organisation: single outlet retailers, small multiples (2-9 outlets) and large multiples (10+ outlets) for 1982, 1987 and 1991.

A first examination of the table reveals that the number of shops belonging to single outlet retailers and small multiple retailers fell between 1982 and 1991, but the number of shops belonging to large multiples increased. The pattern for employment was the same, but turnover increased for all three categories as one might expect as a result of inflation. This leads us to seek a deflator for turnover and to derive a number of ratios and examine their patterns. Obvious ratios are turnover per shop, turnover per employer and, possibly, employees per shop. Tables 4.6 and 4.7 show the first two of these ratios (calculated correctly on unrounded data).

From Table 4.6 we see that turnover per shop in large multiples is roughly five times as high as in small multiples and six or seven times as high as in single outlets. We can also see that turnover per shop doubled between 1982 and 1991 for small and large multiples, but only increased by about 75 per cent for single outlets. Most of the increase for large retailers came between 1982 and 1987, whereas growth was more evenly distributed over the whole period for single outlets and small multiples.

From Table 4.7 we see that turnover per employee in 1991 was over 10 per cent higher for small multiples than in single outlets and about 25 per cent higher in large multiples than in small multiples. Turnover per employee nearly doubled in all three groups between 1971 and 1991. But looking more closely we can see that the increase was slightly smaller for large multiples than single outlets and small multiples (80 per cent compared with 90 per cent roughly). So over time the gap between large and small multiples has closed slightly. We can also see that most of the increase for single outlets was in the earlier period.

Table 4.7 Average turnover[2] per employee[3]: by type of retail organisation

United Kingdom   *£'000*

	Single outlet	Small[4] multiple	Large multiple
1982	25	28	37
1987	38	40	51
1991	47	53	67

Footnotes as for Table 4.5

In cases like this, where two or three related measurements vary over time or between different locations, the ratio between measurements may be more informative than the original data.

Table 4.6 Average turnover per shop: by type of retail organisation

United Kingdom   *£'000*

	Single outlet	Small[4] multiple	Large multiple
1982	98	129	620
1987	138	176	1 000
1991	173	248	1 230

Footnotes as for Table 4.5

Table 4.8 Population change[1]

Thousands

	Population at start of period	Average annual change				
		Live births	Deaths	Net natural change	Other[2]	Overall annual change
Census enumerated						
1901-1911	38,237	1,091	624	467	-82	385
1911-1921	42,082	975	689	286	-92	194
1921-1931	44,027	824	555	268	-67	201
1931-1951	46,038	785	598	188	25	213
Mid-year estimates						
1951-1961	50,290	839	593	246	6	252
1961-1971	52,807	963	639	324	-12	312
1971-1981	55,928	736	666	69	-27	42
1981-1991	56,352	757	655	103	42	145
Mid-year projections[3]						
1991-2001	57,801	786	633	154	53	207
2001-2011	59,719	721	626	95	44	139
2011-2021	61,110	725	644	81	6	87
2021-2031	61,980	710	698	12	0	12

1 See Appendix, Part 1: Population and population projections.
2 Net civilian migration and other adjustments.
3 1991-based projections based on a provisional estimate of the population of the United Kingdom of 57,649 thousand which was subsequently revised.

Source: Office of Population Censuses and Surveys; Government Actuary's Department;
General Register Office (Scotland); General Register Office (Northern Ireland)

Differences

In some cases the difference between two rows or columns may be of particular interest. Consider, for example, Table 4.8 where the 'net natural change' column appears logically as the difference between the entries in the births column and the deaths column for each year. This enables the reader to see immediately that there were more births than deaths each decade this century, but the excess has been declining and is projected to decline still further into the 21st century. The erratic contribution of migration and other adjustments to the overall annual change can also be examined.

Rates per number at risk

It is important that numbers appearing in the same table should be directly comparable. This may involve re-expressing the original numbers as incidence rates per thousand at risk or per hour of exposure to risk. For example, the original data in the first three columns of Table 4.9 might suggest that the RAF is positively careless with its aircraft by comparison with either the Royal Navy or the Army. However, when the number of accidents in each of the services is re-expressed as the number of accidents per 10,000 flying hours, a quite different picture emerges. Measured in these units the RAF's safety record looks altogether more creditable.

Table 4.9   Aircraft accidents involving loss or serious damage: by Service

	Number of accidents			Rate per 10,000 flying hours		
	Royal Navy	Army	Royal Air Force	Royal Navy	Army	Royal Air Force
1975	12	10	23	1.33	0.99	0.45
1980	5	9	24	0.58	1.22	0.50
1985	6	2	18	0.62	0.39	0.37
1990	5	7	16	0.56	1.06	0.32
1995	2	1	8	0.26	0.11	0.22

Source: Defence Statistics 1996

Many accident statistics are more meaningful when quoted as accidents per so many hours exposure to risk (for example accidents to different sorts of sportsmen and accidents associated with different modes of transport). Similarly, absolute numbers of births or deaths are often insufficient to make valid comparisons between regions containing markedly different numbers of women of child bearing age and old people.

Money measured at constant prices

Particularly in times of varying inflation, it may be difficult to discern true patterns of growth or decline in a series of values recorded in money terms. For example, if retail prices have approximately doubled over the last ten years, a shopkeeper whose receipts have risen by only 80 per cent over that period is probably less successful now than he was ten years ago. To reveal true trends in such monetary measurements, the figures should be adjusted to 'constant prices'. This is done by dividing each year's observed value by a deflating factor which takes into account the almost automatic change in the measurement attributable only to inflation. For many series the Retail Price Index will be an appropriate deflating factor: in other cases more specialised index numbers will be used to correct for changes of price in specific categories of goods.

Table 4.10 below shows agricultural output in four categories measured at current prices from 1981-1992 while Table 4.11 shows the same data revalued at constant, 1985 prices.

From Table 4.10 it appears that the value of agricultural output in 1992 was forecast to be roughly 20-30 per cent higher than in 1981-1983 for livestock, livestock products and farm crops, but 70 per cent higher for horticulture. However, when the data are measured at constant prices, as in Table 4.11, it is easier to see that the biggest real increase in agricultural output over this period was in farm crops and that the real value of output of livestock products actually fell by over 10 per cent.

Table 4.11 Agricultural output at constant (1985) prices

United Kingdom				£ million
	Livestock	Livestock products	Farm crops	Horti-culture
Actual				
1981-83	4 250	3 090	2 530	1 240
1988	4 510	2 760	2 980	1 380
1989	4 560	2 690	3 140	1 440
1990	4 660	2 760	3 150	1 390
1991	4 850	2 690	3 170	1 390
Forecast				
1992	4 710	2 670	3 220	1 440

Source: Agriculture in the United Kingdom 1992

Table 4.10 Agricultural output at current prices

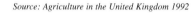

United Kingdom				£ million
	Livestock	Livestock products	Farm crops	Horti-culture
Actual				
1981-83	3 930	2 870	2 820	1 070
1988	4 690	3 030	2 940	1 670
1989	5 050	3 220	3 270	1 810
1990	5 000	3 350	3 400	1 940
1991	5 170	3 310	3 990	1 920
Forecast				
1992	5 220	3 440	3 530	1 820

Source: Agriculture in the United Kingdom 1992

4.3 Presentation of demonstration tables

Analysis of any set of data is likely to involve producing several tables of derived statistics and roughly sketched graphs. If each such table or graph is accompanied by a brief note of the main patterns revealed by this form of presentation, these notes can be used to decide on the final presentation of the data. Generally, only the most important patterns need to be reported and these are usually (though certainly not always) the most striking ones. The derived tables or graphs which best illustrate these patterns should be included in the final report. If tables are used, the numbers in each table must be presented so as to highlight the main patterns in the data.

Demonstration tables should appear in the main body of text (not as annexes or afterthoughts) where they can amplify the argument and encourage the reader to pause momentarily and consider the numbers. There is no place in the main text of a report for 'some figures which may be relevant'. If you do not know precisely what contribution a table makes to the development of an argument, leave it out: the reader is unlikely to find any pattern in a table unless you direct his or her attention to it. You may of course include reference tables separately, probably at the back of the report.

The general rules on structure and layout, described in Chapter 3, apply to all tables including demonstration tables. Briefly these are:

- the table should contain a clear and complete explanation of what the figures represent (kind of objects enumerated and how they have been categorised in the table); the geographical and time coverage; units of measurement and source of the data

- spacing, shading, and horizontal ruled lines should be used to guide the eye, with minimum use of vertical ruled lines

- use a clear typeface with capital letters only for initial letters of first words and proper nouns

- footnotes should be used conservatively but included wherever necessary to prevent misunderstanding or misinterpretation of the data and to give fuller description of compact headings

Thereafter, all the guidelines recommended for constructing effective demonstration tables are designed to highlight patterns and exceptions in the table. These guidelines were formally stated by Ehrenberg[1] who gives the following criterion for a good table:

> 'The patterns and exceptions in a table should be obvious at a glance once the reader has been told what they are.'

It is important to note the last few words of this directive: 'once the reader had been told what they are'. Figures very rarely speak for themselves, and even experienced analysts sometimes pause doubtfully when confronted with a table and an injunction to 'just look at these figures!' When writing for a non-specialist audience it is particularly important that tables (and charts) are used to amplify the text in an obvious and helpful way. The introductory phrase 'As can be seen from Table 1 ....' is likely to appear frequently. The guidelines proposed by Ehrenberg help both to uncover the basic patterns in a set of data and to communicate selected patterns and exceptions to future readers.

---

[1] EHRENBERG, A S C Rudiments of numeracy (with discussion), *Journal of the Royal Statistical Society, Series A: General. 1977, vol 140 para 3 pp 277-297*

4.4 Seven basic rules

These seven basic guidelines are:

1 round all numbers to two effective digits
2 put the numbers to be most often compared with each other in columns rather than rows
3 arrange columns and rows in some natural order or in size order
3a where possible, put big numbers at the top of tables
4 give row and column averages or totals[1] as a focus
5 use layout to guide the eye
6 give a verbal summary of the main points of the table
7 use charts to show relationships

Each guideline will now be considered in more detail, and the effect of implementing it will be demonstrated using Table 4.12 as a simple illustration.

Rule 1: Round to two effective digits

Effective digits are defined as those which vary within a set of data. If the data consist of three digit numbers and all the digits vary, then the rule will be equivalent to rounding to two significant digits. For example, the following data

178   242   575   582   633   750

would be rounded to

180   240   580   580   630   750

If, however, the hundreds digit remained constant throughout then all three digits would be retained.

---

[1] Ehrenberg proposed averages only.

Table 4.12 Motor vehicles currently licensed[1]

Great Britain   *thousands*

Type of vehicle	1953	1963	1973	1983	1993
Private cars[2]	2,446	6,462	11,738	15,543	20,102
Other private & light goods vehicles[2]	516	1,092	1,559	1,692	2,187
Goods vehicles[3][4]	446	535	540	488	428
Motor cycles etc[5]	889	1,546	887	1,290	650
Public transport vehicles[6]	105	86	96	113	107
Agricultural tractors[7]	289	412	373	376	318
Other vehicles[8]	30	88	97	86	55
Crown and exempt vehicles[9]	88	115	137	621 [10]	979
All vehicles	4,809	10,336	15,427	20,209	24,826

1 From 1992 estimates of licensed vehicle stock are taken from the Department of Transport's Statistics Directorate Information database. For years up to 1992 estimates are taken from the Annual Vehicle Census based on the DVLA main vehicle file. The census was taken for the September quarter from 1946-1976 and on 31 December for 1978-1991.
2 For years up to 1990, retrospective counts within these new taxation classes have been estimated.
3 Includes agricultural vans and lorries, showmen's goods vehicles licensed to draw trailers (note 2 applies).
4 Excludes electric goods vehicles which are now exempt from duty.
5 Includes scooters and mopeds.
6 Includes taxis. Prior to 1969, tram cars were included.
7 Includes combine harvesters, mowing machines, digging machines, mobile cranes and works trucks.
8 Includes three-wheelers, showmens' haulage and recovery vehicles.
9 Includes electric vehicle which are now exempt from duty.
10 From 1983 includes old vehicles exempt from tax converted for the first time to the DVLA system.

Source: Department of Transport

Thus the set of numbers

112   126   132   148   165   174

would remain unchanged. This is because, given the range of variation in the data (namely from 112 to 174), the two digits which discriminate between one number and another are the tens and units digits: these are the *effective* digits in this set of data.

Similarly, with four digit numbers the rule might dictate that we retain two, three or four digits depending on how many of the digits change within the set of data. So

1,693   2,382   3,801   4,886   6,181   7,309

would be rounded to

1,700   2,400   3,800   4,900   6,200   7,300

the set

2,114   2,186   2,318   2,468   2,663

would be rounded to

2,110   2,190   2,320   2,470   2,660

and the data

2,211   2,236   2,259   2,274   2,286

would stay unchanged.

The advantage of rounding to two effective digits is that it makes it easy to compare numbers 'at a glance'. Seeing a pattern in a set of data depends upon observing differences or ratios between pairs of numbers: this depends on the reader's ability to do mental arithmetic. Most people can subtract 1,700 from 2,800 mentally, and remember the answer, 1,100, long enough to compare it with the difference between 3,800 and 2,800. Few people, however, can carry out mental arithmetic on three digit numbers. *Without writing anything down*, try to work out whether the difference between 4,886 and 6,181 is bigger or smaller than the difference between 6,181 and 7,309. Now do it with the rounded numbers.

Digits which do not change in a set of data can be mentally filtered out of the calculation, leaving only two digits to be dealt with. Thus, in the second example quoted above, the 2 in the thousands column can be ignored in comparing successive numbers. Using the rounded row,

2,110   2,190   2,320   2,470   2,660

it is easy to verify the difference between successive pairs of numbers increases as we work across the data. This is much less easy to check using the unrounded numbers

2,114   2,186   2,318   2,468   2,663.

The effect of rounding all the numbers in Table 4.12 to two effective digits is as shown in Table 4.13.

Table 4.13 Motor vehicles currently licensed: figures rounded to two effective digits (footnotes omitted)

Great Britain					*thousands*
Type of vehicle	1953	1963	1973	1983	1993
Private cars	2,400	6,500	12,000	16,000	20,000
Other private & light goods vehicles	500	1,100	1,600	1,700	2,200
Goods vehicles	450	540	540	490	430
Motor cycles etc	900	1,500	900	1,300	700
Public transport vehicles	105	86	96	113	107
Agricultural tractors	290	410	370	380	320
Other vehicles	30	88	97	86	55
Crown and exempt vehicles	90	120	140	620	980
All vehicles	5,000	10,000	15,000	20,000	25,000

Source: Department of Transport

The first line in this table illustrates a commonly encountered problem: the number of digits changes - from four to five (2,400 to 20,000). Similar problems occur in other lines. No hard and fast rule is offered here and a common sense approach must be adopted. In Table 4.13 the number of private cars licensed is under 10,000 in 1953 and 1963 and rises to over 20,000 in 1993. The first two entries have therefore been rounded to the nearest hundred and the last three to the nearest thousand. For the 'all vehicles' line, however, most of the entries are over 10,000 and in this case it is probably more helpful to give all the numbers to the nearest thousand, rounding the first entry to 5,000 and the second to 10,000. Similar reasoning has been applied to the entries in other rows of the table, retaining all three digits in the years when the number of public transport vehicles passes 100. Rounding the 'motor cycle etc' line is particularly tricky. On balance rounding to the nearest hundred probably gives the clearest picture of the variation over time. All other rows of the table have been simply rounded to two effective digits.

One snag about rounding to two effective digits is that the figures in the table will not necessarily sum to the printed totals: this can be disconcerting to some readers and, except when writing specifically for an audience whom you know to be knowledgeable about such things, it is wise to put in an explanatory footnote, even though this itself distracts from the directness and simplicity of the presentation. (Something like 'Components may not add to totals because they have been rounded independently' is probably all that is required.) *Do not, on any account, change the figures to make them sum to the totals*.

Having rounded the entries in each row to two effective digits, it is much easier to carry out the mental arithmetic necessary to see the main patterns in the table: from 1953 to 1993 the total number of vehicles licensed increased fivefold, from 5,000 thousand to 25,000 thousand; virtually all of the increase was attributable to the growth in the number of private cars licensed (from 2,400 to 20,000 thousands); other categories showed smaller changes and one category (public transport vehicles) showed little change over the period.

People are frequently reluctant to round numbers to two effective digits on the grounds that they are 'throwing away accuracy'. Certainly accuracy is being sacrificed: what is gained is the ability to communicate a specific quantitative message and that is the purpose of demonstration tables. People do not find numbers easy to remember - but round two-digit numbers are infinitely more memorable than four or five digit numbers.

The magnitude of the error incurred can, of course, always be calculated. If three-digit numbers are rounded to two effective digits, the largest percentage error occurs at the lower end of the range, for example when 104 is rounded down to 100 or 105 is rounded up to 110 there is an error of approximately 5 per cent in either case. At the other end of the range, if 994 is rounded down to 990, the percentage error is about 0.4 per cent. Whether or not the loss of this degree of accuracy is significant depends on the data in question. Assuming that the ultimate objective of any demonstration table is to influence decisions, the acid test must be 'would a different decision be indicated if the precise, unrounded numbers were displayed?' If the answer to this question is 'yes' then clearly it would be wrong to round numbers - but, in such a case, searching questions must be asked about the accuracy of the processes of collection, collation and analysis of the data in order to be completely confident that small differences between comparatively large numbers really exist after allowing for the inevitable errors in these processes.

The other reason that is given for reluctance to round data to two effective digits is that the numbers are then not accurate enough to use in further calculations. This is absolutely true: they are not. However, the numbers in a demonstration table are intended to illustrate patterns and exceptions as effectively as possible: they are not intended for use in further calculations. If you consider it likely that a reader will want to do further calculations based on the numbers in a demonstration table, it is important that he or she should be able to refer to the original data as easily as possible. This can be achieved by giving precise details of the source of the data, if necessary including an address and telephone number for enquiries as a footnote, or, in extreme cases, including a reference table as an appendix to the report. There the reader should find more precisely recorded figures plus an indication of the size of the intrinsic error associated with them, and these are the numbers for use in further calculations.

Rule 2: Put numbers to be compared with each other in columns, not rows

The reasoning behind this rule is very straightforward. Firstly, comparisons often involve mentally subtracting one number from another: we all learned to do subtraction sums by putting the larger number above the smaller and writing the answer underneath; we were trained to subtract in columns. Secondly, it is much easier for the eye to ignore digits which are common to both numbers are vertically aligned. For example, in the case of five-digit numbers where the first digit is common throughout we might have

18,200   14,700   13,500

or   18,200
18,200
14,700
13,500

When the numbers are printed across the page more effort is required to verify that the leading one is common to all numbers and that they all end in two zeros. When they are printed vertically these points are obvious at a glance.

So we have to decide which comparisons will be made most frequently. Clearly this depends on the message(s) which the table illustrates and thus no general rule will apply. However, when data have been recorded over a number of years, time trends in individual categories will often be important. Here this would indicate that the table used as our example be redrawn with rows and columns interchanged, as in Table 4.14.

When rows and columns are interchanged care must be taken to keep the columns evenly spaced. Problems can arise when column headings are of different length but, in such cases, long column headings should be spread over several lines, if necessary breaking long words (eg agricultural) with a hyphen.

Table 4.14 Motor vehicles currently licensed: rows and columns interchanged (footnotes omitted)

Great Britain *thousands*

	Type of vehicle								All vehicles
	Private cars	Other private & light goods vehicles	Goods vehicles	Motor cycles etc	Public transport vehicles	Agricultural tractors	Other vehicles	Crown and exempt vehicles	
1953	2,400	500	450	900	105	290	30	90	5,000
1963	6,500	1,100	540	1,500	86	410	88	120	10,000
1973	12,000	1,600	540	900	96	370	97	140	15,000
1983	16,000	1,700	490	1,300	113	380	86	620	20,000
1993	20,000	2,200	430	700	107	320	55	980	25,000

Source: Department of Transport

Rule 3 and 3a: Arrange columns and rows in some natural order of size: where possible put big numbers at the top

These rules again relate to highlighting patterns and exceptions within the data. If your table includes data about a number of different towns, a natural order might be in decreasing order of population size; if it relates to countries, one natural order might be in decreasing order of wealth, as measured by GDP per head, and another might be in order of physical area. A number of measurements might reasonably be expected to follow such a 'natural' ordering: for example the number of local schools or doctors or policemen or supermarkets might be expected to follow the same order as the population of towns; measurements associated with wealth, such as the number of cars or telephones per head of population, or the amount spent on education per head of population might be expected to follow the same order as GDP per head. Thus when measurements in a table break this 'natural' sequence the 'surprising' order is highlighted.

Where there is no 'natural' order rows and/or columns should be arranged as far as possible in decreasing order of row or column averages. In Table 4.14 this will involve rearranging the columns so that goods vehicles appear after motorcycles. There were more motorcycles than other private and light goods vehicles in 1953 and 1963, but in all later years there were fewer motorcycles. On average over the period shown the motorcycles category was larger, so it appears third. The fourth category will be goods vehicles followed by crown and exempt vehicles. Again the two categories switch places during the period and the ranking has been determined by the average number over the periods shown. Agricultural tractors are sixth, public transport vehicles seventh and 'other' vehicles are last. Any catch-all category such as 'others', 'miscellaneous' or 'unknown' should appear at the end of the data even if it is larger than some of the previous categories. Frequently such a category will be the smallest and so the regular pattern of decrease will be preserved.

In many tables the order in one direction (either rows or columns) will be dictated by the need to follow a chronological order. In Table 4.14, clearly the years determine the ordering of the rows. The guideline which recommends placing big numbers at the top of columns whenever possible is again designed to help the reader carry out subtraction sums easily. it is indeed easier to calculate the difference between 9,700 and 6,100 when they appear like this

> 9,700
> 6,100

than when they appear like this

> 6,100
> 9,700

This suggests that the most recent year should be placed at the top of the table and earlier years beneath it. For some audiences this may be helpful but not for an audience accustomed to using government statistical tables. In virtually all official statistical publications, time is shown as moving either from left to right or from top to bottom, with the most recent year on the extreme right-hand side or at the bottom of a table. Since the aim of a demonstration table is that the reader should find it easy to interpret, it would be thoroughly counter-productive to present the years in an unexpected order.

Where time is neither a row heading nor a column heading then the guidelines should be followed - with, as far as possible, the biggest numbers at the top left hand corner of the table and a general pattern of decrease from that point.

So, rearranging the columns of Table 4.14 we have Table 4.15.

Table 4.15 Motor vehicles currently licensed: columns re-arranged in decreasing order of averages (footnotes omitted)

Great Britain   *thousands*

				Type of vehicle					All vehicles
	Private cars	Other private & light goods vehicles	Motor cycles etc	Goods vehicles	Crown & exempt vehicles	Agricultural tractors	Public transport vehicles	Other vehicles	
1953	2,400	500	450	450	90	290	105	30	5,000
1963	6,500	1,100	540	540	120	410	86	88	10,000
1973	12,000	1,600	540	540	140	370	96	97	15,000
1983	16,000	1,700	490	490	620	380	113	86	20,000
1993	20,000	2,200	430	430	980	320	107	55	25,000

Source: Department of Transport

Three minor points are worth mentioning before leaving the subject of order of rows and columns.

First, for demonstration tables, alphabetic order is not a 'natural' order. In a reference table it is of course helpful to have rows arranged in alphabetic order, so that the reader can look up a desired entry as efficiently as possible; but a demonstration table should be designed to highlight the patterns in the table - not for ease of reference.

Secondly, if you are not confident that your reader is aware of a particular 'natural' ordering (for example, if he or she does not know the populations of the towns or countries included in the table) you should clarify the point by stating explicitly what criterion has determined the row and/or column order.

Finally, if two or three tables with the same row or column headings appear in a report, the same order should be preserved in all tables. If there is no external measure of size to dictate the row and column orders then all the tables should be considered together in order to choose the order which will be most suitable for the group of tables taken together.

Rule 4: Give row and column averages or totals as a focus

The reason for this rule is that averages (where they are appropriate) allow the reader to see an overall pattern by scanning the margins of table. Having absorbed the overall pattern, he or she can then examine the body of the table to see whether the general pattern is repeated in each row of the table, and also to see how much variation there is about each average.

In our example table it would make little sense to work out row averages, that is an average for each year of the private cars and vans, the public transport vehicles, the agricultural tractors and the other vehicles - this would be quite meaningless. The annual total, which is given, is the figure which is appropriate. It is, however, informative to include an average for each type of vehicle calculated for the five years quoted. Some people feel that averages of a time series are of limited value because the focus is on comparisons with the latest or earliest period rather than some indeterminate point in time. However averages allow the reader to examine variations overtime as illustrated here. Table 4.16 includes averages for all columns except Crown and exempt vehicles where a true average cannot be calculated because the coverage of this category increased during the period.

Table 4.16 Motor vehicles currently licensed: column averages included (footnotes omitted)

Great Britain *thousands*

	Type of vehicle								All vehicles
	Private cars	Other private & light goods vehicles	Motor cycles etc	Goods vehicles	Crown & exempt vehicles	Agricultural tractors	Public transport vehicles	Other vehicles	
1953	2,400	500	900	450	90	290	105	30	5,000
1963	6,500	1,100	1,500	540	120	410	86	88	10,000
1973	12,000	1,600	900	540	140	370	96	97	15,000
1983	16,000	1,700	1,300	490	620	380	113	86	20,000
1993	20,000	2,200	700	430	980	320	107	55	25,000
Average	11,000	1,400	1,100	490	..	350	101	71	15,000

Source: Department of Transport

A quick glance at the averages in the bottom row of the table reveals the following points:

- on average, private cars accounted for roughly three-quarters of all vehicles licensed

- only the first three categories (private cars, other private light goods vehicles and motorcycles) have averages of over 1,000 thousand: by comparison with private cars, the remaining categories are very small indeed.

We can now investigate the table year by year to see whether or not these three patterns are true throughout the period. We find:

- it is not until 1973 that private cars assume quite such dominance: in 1953 and 1963 they account for half and two thirds of all vehicles licensed respectively, whereas from 1973 onwards they account for roughly three-quarters of each annual total

- in 1953 and 1963 there were more vehicles in the motorcycles category than other private and light goods vehicles; thereafter the order of categories is the same as that in the average row

- it is only the first category, private cars, which totals over 1,000 thousand, in all years: the second category is 500 in 1953 and over 1000 thereafter; the number of motor cycles fluctuates on either side of 1,000 throughout the period covered by the table.

Finally we can examine individual columns of the table to see how much variation there is about each average and to see whether there is any general relationship between pairs of columns. This reveals:

- considerable variation in the first column: from 2,400 (well below average) to 20,000 (well above the average). The final entry is almost eight times the first entry;

- comparatively little variation about the average in the other columns (some of the apparent growth between 1973 and 1983 in the 'other vehicles' category is attributable to the addition of 170 thousand vehicles which were not included in previous censuses: this is mentioned but not quantified in footnote 10 of the original version of our example table - Table 4.12): except for the third column and the 'other vehicles' column, the average coincides with a value roughly halfway through the period;

- a general decline in the proportion of motorcycles, scooters and mopeds: in 1953 there were just over twice as many private cars as motor cycles; in 1993 there were almost thirty times as many. Viewed differently, in 1953 the motorcycle category accounted for just under 20 per cent of all vehicles, but only 3 per cent in 1993;

- the motorcycles category is unusual in that its numbers declined sharply from 1963 to 1973 and then started to climb, although not to the 1963 level. After 1983, the number fell again to a new low in 1993.

Not all of these observations will appear in the final summary of the table, but using the column averages as a focus has helped to identify a number of patterns and exceptions.

In many tables a choice has to be made between showing row or column averages and showing row or column totals. Averages are sometimes more helpful in interpreting the table than totals. This is because averages are of the same order of magnitude as the entries in the table and can therefore be used as a focus when investigating the patterns of variations within a row or column; by contrast the total of a row with five entries is roughly five times as big as the individual entries.

Rule 5: Use layout to guide the eye

This rule is included here only because it is one of the seven basic rules formulated by Ehrenberg (1977). The recommendations are exactly the same as those discussed in Chapter 3 under the heading 'Spacing of rows and columns and use of ruled lines'. Briefly they are:

• rows and columns within the body of the table should be regularly spaced;

• additional blank spaces should be used to separate a 'total' or 'average' row or column from the body of the table;

• horizontal lines should be included only if they help to guide the reader's eye;

• vertical lines are seldom necessary.

Rule 6: Give a verbal summary of the main points of the table

This is the most important rule. It is essential that a clear and succinct summary should accompany each table. The rule demands that the writer of the report should carefully consider the role of the table. What is the current argument? How does the table contribute to it? What should the reader remember after he or she has scanned the table?

The verbal summary - which may take the form of bullet points - should contain only a few key points (say three or four); it should never be a blow-by-blow account of each entry in the table. It is usually a mistake to include explanations about changes of

definition in the verbal summary. In general these belong as footnotes although there may by occasions when it is desirable to emphasise that an apparently interesting change in a trend is due to a change in definition: if this is thought necessary the explanation should appear at the end of the verbal summary rather than as the first or second point.

Clearly the emphasis given to patterns and trends in any table will be dictated by the role of the table in a particular report and on the expected readership. This is inevitably a subjective decision and so it is important to consider the table carefully so as to ensure that your summary is an honest exposition of the data. For example, in Table 4.16, even if your audience is likely to be particularly interested in public transport vehicles, it would not be appropriate to dwell at length on the changes in that category of vehicle (from 105 to 86 and back to 107 thousand) without pointing out the much more dramatic change in the number of private cars licensed.

On the assumption that Table 4.16 is intended for a general report on trends in the numbers of licensed vehicles, the verbal summary below would be suitable.

Verbal summary to Table 4.16

Table 4.16 shows that the number of licensed vehicles in Great Britain increased fivefold between 1953 and 1993 from 5 million to 25 million. This increase was almost entirely due to the growth in private cars. By 1973, three quarters of the licensed vehicles were private cars compared with half in 1953. Most of this growth took place in the 1950s and 1960s. Although they make a comparatively small contribution to the total number of vehicles, there has been a more than tenfold increase in the numbers in 'Crown and exempt' category, from 90 thousand to 980 thousand, over the period. On the other hand, the number of motorcycles has fluctuated over the period, falling in the most recent period (to 0.7 million in 1993).

This gives a clear, concise summary of the main patterns in the table. The table is specifically referred to by its number in order to link it in with the text; the numbers on which the summary statements are based are quoted (... 'the number increased in the 1970s reaching 1.3 million in 1963'); and no attempt is made to say something about each category in turn. Essentially the verbal summary is well designed to help an interested but non-specialist audience appreciate the main patterns in the table.

It is a useful discipline to write the verbal summary *before* deciding on the final design of the table to include in a report. This will help to ensure that the figures most frequently compared with each other are shown in columns as well as ensuring that the statistics quoted in the table are those which most effectively illustrate the main patterns in the data.

Writing a verbal summary will also help to decide how may separate categories should be shown in the final table. Many tables include a final category labelled 'others' or 'miscellaneous'. In Table 4.16 there are a number of minor categories as well as the 'other vehicles' one. Of these only the 'crown and exempt' category is commented on in the verbal summary and it is therefore reasonable to consider absorbing the three other small categories into an enlarged 'others' category.

It will not always be immediately obvious where minor categories can be safely amalgamated: their pattern of growth or decline must be compared with trends in other columns of the table - best done by applying rules 1 to 4 to the whole table - in order to establish whether or not a comparatively small category merits specific mention in the verbal summary. In Table 4.16 the more than ten fold growth in Crown and exempt vehicles was included for this reason. When nothing particularly noteworthy has been observed in minor categories, the final table will be more compact and the main patterns clearer if these small categories are absorbed into an expanded 'others' category. Since the proposed verbal summary makes no mention of either public transport vehicles or agricultural, the amalgamation seems sensible and the final table is shown at Table 4.17.

In Table 4.17 the commas separating thousands and hundreds have been dropped and the footnotes have been simplified with regard to the content of the taxation classes to suit the general reader, but made more helpful with regard to the significant discontinuities in the data.

Table 4.17 Motor vehicles currently licensed[1]: simplified version of table with simplified footnotes

Great Britain   *thousands*

	Private cars	Other private & light goods vehicles	Motor cycles etc	Goods vehicles	Crown & exempt vehicles	Other vehicles	All vehicles[2]
1953	2 400	500	900	450	90	420	5 000
1963	6 500	1 100	1 500	540	120	580	10 000
1973	12 000	1 600	900	540	140	560	15 000
1983	16 000	1 700	1 300	490	620[3]	580	20 000
1993[4]	20 000	2 200	700	430	980	480	25 000
Average	11 000	1 400	1 100	490	..[3]	530	15 000

1 Taxation classes changed in 1990. Figures for earlier years have been estimated on the new basis. All classes except motor cycles were affected, particularly the private cars and other private and light goods vehicles classes.
2 Components may not add due to rounding.
3 Includes some vehicles previously exempt from tax and not included in previous censuses. There were 170,000 in 1981. A meaningful average cannot be calculated because of the discontinuity.
4 Method of estimation changes in 1992. Estimates for 1992 were 1 per cent lower on the new basis than on the old basis.

Source: Department of Transport

Rule 7: Use charts to show relationships

Charts are excellent for conveying broad trends and relationships: 'the rate of growth increased sharply in 1994'; or 'in 1981 categories A and B were roughly the same size: by 1991 category B completely dominated category A'; or 'the percentage of liquid steel made using the open hearth process decreased markedly during the 1960s'; or series 'A and B declined at similar rates during the 1980s: during the same period, series C grew rapidly, overtaking both series A and B by 1986'.

Messages like these can be illustrated effectively using charts: a line graph for the first; grouped bar charts for the second; pie charts for the third and three line graphs for the fourth. But in all cases the reader will be left with a general impression rather than a memory of any specific amounts. This may be completely appropriate. If you find that your verbal summary includes a number of qualitative statements that do not depend on particular numerical values, you should pause and consider whether the main patterns in your data might be communicated more effectively using a chart rather than a table. Chapters 6 and 7 are devoted to the subject of statistical charts.

However, even where general statements are included in the verbal summary, the data may not lend themselves to a graphical presentation. For example, consider the verbal summary to Table 4.16. The four points singled out for comment are:

1   the increase in the total number of licensed vehicles
2   the increase in the number of private cars
3   the marked increase in the comparatively small category, 'Crown and exempt'
4   the fluctuating numbers of motorcycles.

It is extremely difficult to devise a single chart which illustrates these four points as effectively as the table. If the total number of vehicles is represented by a series of vertical bars (as in Figure 4.1), the overall pattern of growth can be seen. But any attempt to subdivide these bars into six slices of changing sizes as in Figure 4.2 leads to a muddled impression, and the changes in the number of motorcycles and growth in Crown and exempt vehicles are almost impossible to distinguish. If the categories are plotted as line graphs against time as in Figure 4.3, we find that the scale needed to include the private vehicles numbers on a compact chart, reduces the other lines to low horizontal wiggles. It is hard to distinguish motor cycles and crown and exempt from the other three lines.

Figure 4.1 Motor vehicles currently licensed

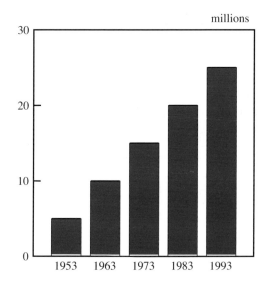

Figure 4.2 Motor vehicles currently licensed

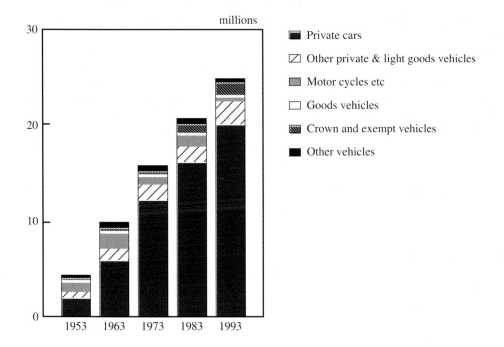

Figure 4.3 Motor vehicles currently licensed

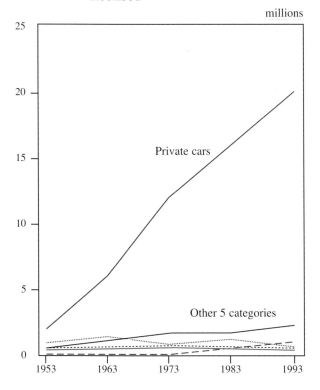

These arguments are not meant to deny the value of statistical graphs and diagrams: for some data displays good charts are much more effective than tables. However, some sets of data are unsuitable for graphical presentation because of the scales involved. Charts are also likely to be less effective than well designed tables if the reader is expected to notice or remembers specific numbers.

Finally it is worth repeating that demonstration tables should appear in the body of a report. Very few readers will take the trouble to flip backwards and forwards through a report looking for tables in an annex - and so very few readers will pay any attention to the quantitative basis of your argument. (If the figures don't matter, leave them out.) This means that tables should be compact enough to be included in the text. They should also appear as close as possible to their verbal summary, either on the same page or on a facing page so that the table can be consulted easily while the verbal summary is being read. This proximity is much easier to achieve if you use several small demonstration tables, each illustrating only two or three points rather than a single large table.

Occasionally, when the message to be conveyed is not a simple one, it is right to venture beyond the rules and guidelines given in this chapter and to *dramatise* a table by turning it into a picture. An example is the comparison of unemployment statistics.

Most European countries regularly collect information about the working life of their population in a survey of households called the Labour Force Survey. This allows statisticians to measure employment on a consistent basis across the European Union using a definition decided by the International Labour Office (ILO). This ILO measure can be compared with other national definitions using information collected in the survey. UK statisticians in the Employment Department produced Table 4.18 to compare this ILO measure of unemployment with the number of people claiming unemployment related benefits, often called the monthly 'headline unemployment' figures. There are a number of interesting patterns in the data but none which stands out clearly in the table.

Readers have to do a certain amount of mental arithmetic to check they understand which rows add to which totals before they can begin to spot the key points.

Consider instead Figure 4.4 which brings the figures to life. It shows clearly that the total number of unemployed was very similar on both measures in Spring 1993. We can see that there was a common core of 1.86 million claimants who were also ILO unemployed, but that there were substantial non-overlapping components in each measure. (Nearly a million claimants were not ILO unemployed and vice versa.) The diagram for men shows visually that there is a substantial overlap between the two measures of unemployment, but that there are more men claimants than are deemed to be ILO unemployed. For women the reverse is clearly true and there is far less overlap between the measures. The chart reveals clearly several interesting points which are difficult to pick out from a standard table. Unfortunately there are some problems with the scaling of the boxes and we are unable to award a ☑ .

Table 4.18 ILO measure of unemployment compared with the monthly claimant count

Great Britain, millions[a]

	Spring 1993			Spring 1992			Change since spring 1992		
	All	Men	Women	All	Men	Women	All	Men	Women
Total ILO unemployed (available for work and looked for work in the last four weeks)[b]*	2.80	1.90	0.90	2.65	1.79	0.86	0.16	0.12	0.04
of which:									
Not in the claimant count	0.95	0.43	0.51	0.89	0.39	0.50	0.06	0.05	0.01
Claimants **	1.86	1.47	0.39	1.76	1.40	0.36	0.10	0.07	0.02
Claimants ** not unemployed[c]	1.00	0.73	0.27	0.85	0.61	0.24	0.15	0.12	0.03
of which:									
not seeking in last four weeks or not available (inactive)[d]	0.62	0.44	0.18	0.53	0.38	0.16	0.09	0.06	0.03
Employed									0.00
Total claimant count	2.86	2.20	0.66	2.61	2.01	0.61	0.25	0.19	0.05

*   See *technical note* for detailed definition.
** These figures are derived with reference to both the claimant count and the LFS results. See *technical note* for details of the method used.
a   Figures may not appear to add because of rounding.
b   Of which, in spring 1993, 110,000 were aged under 18, compared with 120,000 in spring 1992
c   Not unemployed on the ILO definition.
d   People not in work, nor unemployed on the ILO definition.

Figure 4.4 ILO unemployed compared with claimant unemployed

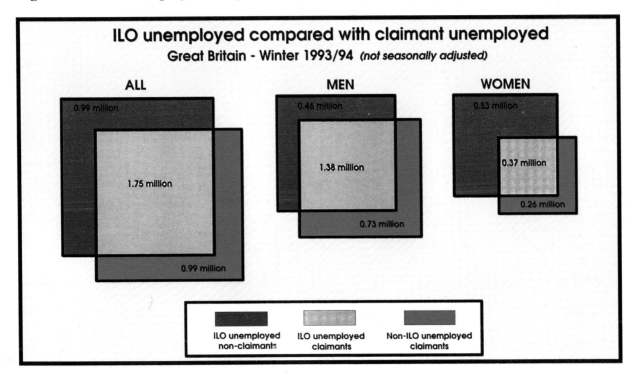

No rules can be given on how to dramatise tables; it is largely a matter of improvisational flair. What can be said, however, is that such unorthodox methods should only be used in particular cases where they offer clear advantage over orthodox tables (or charts) and should always present the message without distortion. They are likely to be much more effective when used with full knowledge of the rules and guidelines set out in this and other chapters.

5 Reference Tables

5.1 Introduction

The criterion for a good reference table is extremely simple: it should be easy to use. This means that the reader should have no difficulty in identifying the row and/or the column which is wanted, no difficulty in reading along a row or down a column to the appropriate entry, no difficulty in finding the exact definition of any entry and no doubt about where, when and how the data were collected.

Many of the guidelines which are important in designing good reference tables have already been discussed in Chapter 3. Points of style which need special consideration for reference tables are:

- which categories to show separately
- which factor to show in rows and which in columns
- ordering of rows and columns
- layout of large tables
- treatment of errors
- footnotes and general explanatory notes covering more than one table
- key to abbreviations and special symbols used
- explanation of how the data were collected
- commentary on table.

5.2 Choice of categories to be tabulated

In constructing any table the compiler must decide whether to tabulate each category for which separate data are available or whether to combine some of them. This task sometimes poses a severe test of statistical techniques and judgement, and advice on such matters is largely outside the scope of this book. There are however a few simple guidelines for general application.

In reference tables, the category set should be as full as is likely to be required by users, consistent with cost constraints and reasonable size of tables. Sub-totals should be provided where appropriate and, if the primary data allow it, the chosen categories should conform to a standard set, such as those in the Standard Industrial Classification (SIC).

The advantage of using a standard set of categories is that the data can then be compared directly with other data collected according to the same classification. This is an important point as data in reference tables will frequently be used for further analysis involving wider comparisons.

Unfortunately not all standard sets of categories are compatible with each other: for example some, but not all, of the categories defined in the Standard Industrial Classification coincide with those used in the Standard International Trade Classification (SITC). (At the highest level aggregation, the SIC section 'Manufacturing' includes food, drink and tobacco manufacturing industries; in the SITC the section 'Manufactures' omits food, drink and tobacco. At more detailed levels of aggregation there are many minor differences in definition between categories which at first sight appear the same.)

Similarly if you categorise data on the social class of respondents according to the Registrar General's socio-economic grouping, your data will not be directly comparable with tables in which respondents are classified according to the Institute of Practitioners in Advertising (IPA) categories. And vice versa.

In cases like this, the only sensible approach is to consider carefully the likely needs of users of your table(s). If possible ask them what analyses they carry out on the data and what other data sources they use. Discuss with them the advantages and disadvantages of different choices of categories and try to select a set of categories which is helpful to as many users as possible. (If this sounds unrealistic because you cannot identify the users of your data, it is reasonable to ask why the data are being published at all.)

Clearly one of the most likely comparisons is with the same data collected in previous years. If a reference table is compiled regularly, the choice of categories to show separately will be made almost automatically: the same as before. This is usually the correct decision, but not always. If the data are to be used as the basis for new comparisons it may be desirable to change the categories shown,

Table 5.1 Gross domestic product: by category of expenditure

£ million, 1990 prices

	Consumers' expenditure+[1]	Central government	Local authorities	Total	Gross domestic fixed capital formation+	Value of physical increase in stocks and work in progress+[2]	Total domestic expenditure [1]	Exports of goods and services+	Total final expenditure[1]	less Imports of goods and services+	Statistical discrepancy (expenditure adjustment)[4]	Gross domestic product at market prices[1]	less Factor cost adjustment[3]	Gross domestic product at factor cost[4]
	CCBH	DJDK	DJDL	DJCZ	DFDM	DHBK	DIEL	DJCV	DJDA	DJCY	GIXS	CAOO	DJCU	CAOP
1989	345 406	68 836	41 303	110 139	110 503	3 669	569 717	126 836	696 553	147 615	–	548 938	72 712	476 226
1990	347 527	70 108	42 826	112 934	106 776	−1 118	566 119	133 284	699 403	148 285	–	551 118	72 232	478 886
1991	339 993	71 950	43 847	115 797	96 265	−4 722	547 333	132 114	679 447	140 248	−430	538 769	71 049	467 720
1992	339 610	72 678	43 957	116 635	94 741	−1 773	549 213	135 547	684 760	148 271	−668	535 821	70 250	465 571
1993	..	..	..	..	..	..	..	..	..	..	..	..	..	474 665
	CAAB	DIAV	DIAW	DIAT	DECU	DGBA	DIAY	DJDG	DIAU	DJDJ			DIAS	
1988 Q3	84 690	16 592	10 276	26 868	26 288	1 322	139 168	30 707	169 875	34 719	–	135 156	17 943	117 213
Q4	85 562	17 046	10 215	27 261	26 629	3 469	142 921	30 032	172 953	36 602	–	136 351	18 248	118 103
1989 Q1	85 847	17 023	10 259	27 282	28 343	953	142 425	31 496	173 921	37 133	–	136 788	18 120	118 668
Q2	86 472	16 925	10 260	27 185	27 551	1 485	142 693	31 000	173 693	36 468	–	137 225	18 327	118 898
Q3	86 243	17 545	10 378	27 923	27 449	1 345	142 960	31 759	174 719	37 350	–	137 369	18 239	119 130
Q4	86 844	17 343	10 406	27 749	27 160	−114	141 639	32 581	174 220	36 664	–	137 556	18 026	119 530
1990 Q1	86 992	17 547	10 443	27 990	27 628	27	142 637	33 259	175 896	37 630	–	138 266	18 112	120 154
Q2	87 409	17 484	10 673	28 157	27 124	547	143 237	33 264	176 501	37 487	–	139 014	18 433	120 581
Q3	86 778	17 448	10 824	28 272	26 397	−133	141 314	33 110	174 424	36 881	–	137 543	18 029	119 514
Q4	86 348	17 629	10 886	28 515	25 627	−1 559	138 931	33 651	172 582	36 287	–	136 295	17 658	118 637
1991 Q1	85 834	17 902	10 917	28 819	24 669	−1 078	138 244	31 932	170 176	34 908	−84	135 184	17 707	117 477
Q2	84 806	18 086	10 944	29 030	24 063	−1 692	136 207	33 159	169 366	34 726	−104	134 536	17 657	116 879
Q3	84 712	18 039	10 990	29 029	23 750	−1 535	135 956	33 475	169 431	34 923	−117	134 391	17 715	116 676
Q4	84 641	17 923	10 996	28 919	23 783	−417	136 926	33 548	170 474	35 691	−125	134 658	17 970	116 688
1992 Q1	84 227	18 223	10 983	29 206	23 817	−858	136 392	33 460	169 852	36 273	−145	133 434	17 443	115 991
Q2	84 717	18 525	10 956	29 481	23 548	−733	137 013	34 078	171 091	37 262	−160	133 669	17 589	116 080
Q3	85 089	17 905	11 002	28 907	23 450	274	137 720	33 819	171 539	37 222	−175	134 142	17 579	116 563
Q4	85 577	18 025	11 016	29 041	23 926	−456	138 088	34 190	172 278	37 514	−188	134 576	17 639	116 937
1993 Q1	85 981	17 871	11 048	28 919	24 032	−766	138 166	35 082	173 248	37 766	−155	135 327	17 735	117 592
Q2	86 432	18 765	10 406	29 171	23 386	241	139 230	34 597	173 827	37 424	−153	136 250	18 015	118 235
Q3	87 268	18 699	10 444	29 143	23 560	−369	139 602	35 591	175 193	37 900	−151	137 142	18 140	119 002
Q4	..	..	..	..	..	..	..	..	..	..	..	..	..	119 836

Percentage change, quarter on corresponding quarter of previous year

1988 Q3	7.7	−1.7	0.4	−0.9	10.4		6.9	−0.5	5.5	8.8		4.7	5.0	4.6
Q4	7.1	1.7	−0.2	1.0	7.0		8.5	1.8	7.3	19.2		4.5	5.3	4.4
1989 Q1	5.0	−0.2	−0.3	−0.2	10.7		5.5	5.4	5.5	14.7		3.2	3.1	3.3
Q2	4.7	0.2	0.2	0.2	5.1		4.6	1.4	4.0	8.0		3.0	3.6	2.9
Q3	1.8	5.7	1.0	3.9	4.4		2.7	3.4	2.9	7.6		1.6	1.6	1.6
Q4	1.5	1.7	1.9	1.8	2.0		−0.9	8.5	0.7	0.2		0.9	−1.2	1.2
1990 Q1	1.3	3.1	1.8	2.6	−2.5		0.1	5.6	1.1	1.3		1.1	0.0	1.3
Q2	1.1	3.3	4.0	3.6	−1.5		0.4	7.3	1.6	2.8		1.3	0.6	1.4
Q3	0.6	−0.6	4.3	1.2	−3.8		−1.2	4.3	−0.2	−1.3		0.1	−1.2	0.3
Q4	−0.6	1.6	4.6	2.8	−5.6		−1.9	3.3	−0.9	−1.0		−0.9	−2.0	−0.7
1991 Q1	−1.3	2.0	4.5	3.0	−10.7		−3.1	−4.0	−3.3	−7.2		−2.2	−2.2	−2.2
Q2	−3.0	3.4	2.5	3.1	−11.3		−4.9	−0.3	−4.0	−7.4		−3.2	−4.2	−3.1
Q3	−2.4	3.4	1.5	2.7	−10.0		−3.8	1.1	−2.9	−5.3		−2.3	−1.7	−2.4
Q4	−2.0	1.7	1.0	1.4	−7.2		−1.4	−0.3	−1.2	−1.6		−1.2	1.8	−1.6
1992 Q1	−1.9	1.8	0.6	1.3	−3.5		−1.3	4.8	−0.2	3.9		−1.3	−1.5	−1.3
Q2	−0.1	2.4	0.1	1.6	−2.1		0.6	2.8	1.0	7.3		−0.6	−0.4	−0.7
Q3	0.4	−0.7	0.1	−0.4	−1.3		1.3	1.0	1.2	6.6		−0.2	−0.8	−0.1
Q4	1.1	0.6	0.2	0.4	0.6		0.8	1.9	1.1	5.1		−0.1	−1.8	0.2
1993 Q1	2.1	−1.9	0.6	−1.0	0.9		1.3	4.8	2.0	4.1		1.4	1.7	1.4
Q2	2.0	1.3	−5.0	−1.1	−0.7		1.6	1.5	1.6	0.4		1.9	2.4	1.9
Q3	2.6	4.4	−5.1	0.8	0.5		1.4	5.2	2.1	1.8		2.2	3.2	2.1
Q4	..	..	..	..	..		..	..	..	..		..	..	2.5

1 These series are affected by the abolition of domestic rates and the introduction of the Community Charge. For details, see notes in the UK National Accounts article in the latest edition of *UK Economic Accounts*.

2 Includes quarterly alignment adjustment. For explanation see notes in the UK National Accounts article in the latest edition of *UK Economic Accounts*.

3 Represents taxes on expenditure less subsidies, both valued at 1990 prices.

4 GDP is estimated in seasonally adjusted form only. Therefore whilst seasonally and unadjusted versions exist of the residual error, the attribution of statistical discrepancies to the expenditure-based and income-based estimates can be made only in seasonally adjusted form.

Source: Central Statistical Office

but this decision can only be made in consultation with the main users of the data.

Having chosen categories to be represented explicitly in the table, one is usually left with a remainder or 'other' category. Whenever possible the 'other' category should be smaller than the smallest category shown separately. In this flawed world there is often also an 'unknown' category. It is bad practice to combine these two into an 'other and unknown' category except when the combined category is small.

5.3 Which factors to put in rows and which in columns

It is easier to scan down a column in search of a particular item than to search for it along a row. This means that, in designing a reference table, you must put yourself in the user's position to decide how the table should be oriented.

Consider for example, Table 5.1 giving gross domestic product by category of expenditure from 1989 to 1993. A user of this table is unlikely to extract a single number from the table. He or she is most likely to want to study the trend over time within one category. Thus, having found the column (or row, if the table had been produced the other way round) of interest, the user will typically copy down several numbers from this column. This extraction is easier if time is shown vertically, as in Table 5.1.

In Table 5.2 (from *Employment Gazette* December 1993) which records unemployment by age and duration, users are likely to compare the distribution of durations of unemployment for different age groups. This entails first identifying the age group to be studied (column headings), then for each age group in turn recording the number of people who had been unemployed for under one week, over one and up to two weeks, and so on: in other words finding and copying down a number of items from the same column. Thus, for each search along a row

Table 5.2 Unemployment by age and duration, October 1993

GREAT BRITAIN Duration of unemployment in weeks	AGE GROUPS												
	Under 18	18	19	20-24	25-29	30-34	35-39	40-44	45-49	50-45	55-59	60 and over	All ages
MALE													
One or less	711	2,690	2,242	12,230	9,269	6,605	4,819	4,080	3,859	3,182	2,650	1,001	53,338
Over 1 and up to 2	714	2,763	2,467	14,167	10,744	7,440	5,698	5,103	4,964	4,693	3,655	1,690	64,098
2 4	1,317	5,509	4,377	22,354	15,928	11,365	8,159	6,941	6,398	5,462	4,393	1,719	93,922
4 6	1,173	7,689	4,607	19,896	14,123	10,346	7,442	6,419	5,818	4,909	3,993	1,651	88,066
6 8	862	3,431	3,062	15,554	11,784	8,534	6,258	5,343	4,983	4,875	3,819	1,570	70,075
8 13	1,667	7,601	7,065	36,012	26,765	19,165	13,826	11,728	10,825	9,624	8,060	3,367	155,705
13 26	1,967	13,705	13,209	70,181	51,845	37,874	27,422	23,171	21,679	20,449	17,625	7,857	306,984
26 39	754	8,124	7,651	39,045	36,020	27,913	20,719	17,832	16,602	16,804	16,619	8,706	216,789
39 52	265	4,791	6,426	31,184	29,005	22,124	16,363	13,775	12,819	12,217	11,686	6,556	167,211
52 65	148	814	9,244	28,239	25,549	19,153	14,175	11,603	10,568	10,255	9,719	3,357	142,824
65 78	46	354	5,102	19,271	17,880	13,973	10,434	8,524	7,482	6,873	6,963	967	97,869
78 104	11	221	5,485	28,270	29,754	23,943	18,040	15,215	13,444	11,983	11,764	1,312	159,442
104 156	0	49	258	40,682	44,295	36,801	27,902	23,431	20,716	17,474	16,187	1,498	229,293
156 208	0	0	7	13,295	18,969	16,281	12,099	10,402	8,919	7,596	6,973	533	95,074
208 260	0	0	0	3,742	6,934	6,299	4,590	4,067	3,762	3,693		232	36,639
Over 260	0	0	0	1,854	7,687	9,585	9,134	9,446	9,675	10,461	20,614	753	79,209
All	9,635	57,741	71,202	395,976	356,551	277,401	207,080	177,080	162,513	150,177	148,413	42,769	2,056,538
FEMALE													
One or less	471	1,865	1,502	5,966	3,357	1,868	1,391	1,414	1,570	1,129	797	4	21,334
Over 1 and up to 2	541	1,927	1,533	7,014	4,293	2,483	1,783	1,744	1,945	1,530	1,037	6	25,836
2 4	954	3,938	2,811	10,733	6,060	3,497	2,368	2,469	2,587	1,901	1,308	7	38,633
4 6	963	6,654	3,089	9,751	5,809	3,428	2,649	2,620	2,375	1,828	1,275	5	40,446
6 8	625	2,248	2,027	7,648	4,773	2,824	2,131	2,135	2,267	1,880	1,336	6	29,900
8 13	1,293	4,526	4,323	16,585	9,982	5,984	4,273	4,326	4,772	3,728	2,686	18	62,496
13 26	1,493	7,659	7,885	31,058	18,745	11,554	8,067	8,269	9,076	7,301	5,829	38	116,974
26 39	514	3,943	3,778	14,593	12,383	8,096	5,794	5,923	6,781	5,989	5,487	40	73,321
39 52	198	2,077	3,032	10,643	9,024	5,828	3,904	4,115	4,695	4,028	3,625	31	51,200
52 65	127	413	4,203	8,920	6,207	3,981	3,078	3,257	3,712	3,397	3,096	17	40,408
65 78	39	173	1,983	5,715	3,261	2,105	1,766	1,922	2,299	2,201	1,967	12	23,443
78 104	13	97	1,954	6,943	4,613	2,992	2,466	2,914	3,557	3,323	3,356	18	32,246
104 156	0	33	102	9,787	5,946	3,892	3,158	3,844	4,761	4,440	4,132	22	40,117
156 208	0	0	3	2,711	2,363	1,513	1,217	1,639	2,079	1,978	1,816	16	15,335
208 260	0	0	0	753	880	624	419	636	841	956	1,059	12	6,180
Over 260	0	0	0	476	1,401	1,139	925	1,126	1,840	2,579	6,742	119	16,347
All	7,231	35,553	38,225	149,296	99,097	61,808	45,389	48,353	55,157	48,188	45,548	371	634,216

which is comparatively slow, the user will carry out a number of vertical searches which are quicker and easier.

A second principle is that it is generally better to have figures of about the same size arranged in columns rather than rows, because this results in a neater and easier to use table.

Occasionally these two principles conflict but more often than not they reinforce one another. For instance, when time is one of the factors in a table, both principles call for the time categories to be arranged vertically. Demonstration of the improvement brought about by such an arrangement is provided by Tables 3.9, 3.10 and 3.11 (pages 18,19 and 20).

5.4 Order of tabulation of categories

Some categories, such as time periods, have their own natural order but others leave you free to choose the order you put them in.

In tables, use the relevant standard order, for example, alphabetical order or that of the Standard Industrial Classification. If no standard order exists, decide your own order and stick to it in future tables so that related figures from tables for different years can be easily compared or extracted.

It is generally good practice to put the 'unknown' category last and the 'other' category second last if these two categories appear in a table.

As usual, the important principle is to think carefully about the order of categories and make a deliberate choice with the interest of users in mind.

Total and sub-total columns should be placed to the right of their components as recommended in Chapter 3. In a complex table it may be helpful to number the columns and expand the headings of any total and subtotal columns to show which components are included eg 'Total (cols 3+6+9)' as in Table 3.11.

5.5 Layout of large tables

In general, reference tables should be printed vertically on the page rather than horizontally. It is often very simple to redesign a table so that it can be printed vertically. However, if a number of reference tables are all wide and shallow, it will be convenient for the whole set to be printed horizontally. See, for example, Table 5.3 which is one of a large set of tables giving statistics on tourism in 1989, 1990 and 1991. Tables set horizontally should always be printed the same way round, with the top on the left for the convenience of the reader.

An alternative layout for very wide tables is to extend them across two pages. When this is done the title on the initial page should be repeated on the continuation page followed by '(*continued*)' in italic type.

A double page spread may be the best way to present a table spanning two pages but care needs to be taken with this form of table because it requires precise alignment of the two pages by the printer - not always easy to achieve even today.

Where a table spans two pages it may be useful to repeat the row headings at the righthand side of the table. This is particularly desirable when there are large numbers of rows close together and it may be difficult to read across all the columns accurately.

5.6 Errors: indication of size of intrinsic error

Figures in tables are rarely accurate to the last digit. Almost every figure in almost every table has associated with it a range of probable error. Errors may arise at each stage of the process of data collection and storage: the original data may be incomplete, either by accident (non-return) or design (if the data are collected from a sample survey); the data may have been inaccurately recorded or transcribed; some data may have been assigned to incorrect categories; and simple arithmetic mistakes may have occurred in calculating totals or percentages. These factors all contribute to the intrinsic lack of accuracy of published data.

OVERSEAS VISITORS TO THE UK

TABLE 4

Table 5.3 Extract from Business Monitor MA6 1991

(thousands)

NUMBERS OF VISITS BY PERMANENT RESIDENCE, MAIN PURPOSE OF VISIT AND MODE OF TRAVEL

	AIR			SEA			ALL ROUTES		
	1989	1990	1991	1989	1990	1991	1989	1990	1991
NORTH AMERICA									
Holiday	1,305	1,383	997	381	477	199	1,686	1,861	1,196
(of which inclusive tour)	(288)	(344)	(182)	(147)	(192)	(76)	(435)	(536)	(258)
Business	609	656	560	22	20	14	631	675	574
*VFR	691	729	671	63	66	24	754	795	695
Miscellaneous	360	372	277	51	47	29	411	418	306
All visits	**2,964**	**3,140**	**2,505**	**517**	**610**	**266**	**3,481**	**3,749**	**2,772**
EUROPEAN COMMUNITY									
Holiday	1,229	1,367	1,331	2,235	2,027	2,406	3,463	3,394	3,737
(of which inclusive tour)	(277)	(311)	(309)	(615)	(662)	(723)	(892)	(973)	(1,031)
Business	1,952	2,017	1,926	620	535	580	2,572	2,553	2,506
*VFR	924	1,043	1,020	880	775	834	1,804	1,818	1,854
Miscellaneous	579	596	573	542	498	551	1,121	1,094	1,125
All visits	**4,684**	**5,023**	**4,850**	**4,276**	**3,835**	**4,372**	**8,960**	**8,858**	**9,222**
OTHER WESTERN EUROPE									
Holiday	515	568	518	161	134	171	677	702	689
(of which inclusive tour)	(240)	(241)	(225)	(75)	(46)	(82)	(314)	(287)	(308)
Business	486	547	418	40	24	21	526	571	440
*VFR	220	248	259	67	33	30	287	280	289
Miscellaneous	168	190	201	71	44	40	239	234	241
All visits	**1,389**	**1,552**	**1,396**	**339**	**235**	**262**	**1,728**	**1,787**	**1,658**
OTHER AREAS									
Holiday	1,221	1,379	1,042	240	364	277	1,460	1,743	1,320
(of which inclusive tour)	(290)	(308)	(219)	(90)	(158)	(111)	(380)	(466)	(330)
Business	598	651	583	36	44	30	634	695	613
*VFR	589	649	596	63	74	63	651	723	659
Miscellaneous	385	420	369	38	46	53	423	465	421
All visits	**2,792**	**3,099**	**2,590**	**377**	**527**	**423**	**3,168**	**3,627**	**3,013**
ALL AREAS									
Holiday	4,270	4,697	3,888	3,016	3,003	3,054	7,286	7,700	6,942
(of which inclusive tour)	(1,095)	(1,204)	(935)	(927)	(1,058)	(992)	(2,021)	(2,262)	(1,927)
Business	3,645	3,871	3,487	717	623	646	4,363	4,494	4,133
*VFR	2,423	2,669	2,546	1,074	947	952	3,497	3,616	3,498
Miscellaneous	1,492	1,577	1,420	701	634	672	2,193	2,211	2,092
All visits	**11,829**	**12,814**	**11,341**	**5,509**	**5,207**	**5,323**	**17,338**	**18,021**	**16,664**
(of which excursionists) (1)	(614)	(374)	(570)	(415)	(509)	(438)	(1,029)	(883)	(1,009)

* VFR = Visiting Friends or Relatives.
(1) See note 2 on page 26

Table users should ideally be given some idea of the size of intrinsic errors in the figures, and yet the great majority of tables contain no such indication.

Why is this? There are two main problems. First the size of the error may not be known, even to the person who collected or compiled the data. This is so, for example, when the data collection system relies heavily on the truthfulness and reliability of people. Secondly, even when the probable size of error is known (for example, the results from a properly designed and well-regulated sample survey) there is no convenient and simple means of showing it. Broadly speaking, you can do one of four things when compiling tables:

1 ignore the existence of intrinsic errors altogether. This is often done by describing the figures in the tables as numbers 'recorded' without any reference to how closely these correspond to the numbers of *actual* objects, events, etc which they represent

2 make a general disclaimer covering the whole set of tables, saying, for example, that figures are not necessarily accurate to the last digit shown

3 indicate the size of errors implicitly by rounding figures to an appropriate degree

4 indicate the size of errors explicitly, by showing confidence intervals[1] or by giving a 'quality label' to each figure or set of figures (eg label A could indicate ranges of error less than ±1 per cent, B those between ±1 per cent and ±10 per cent and C those greater than ±10 per cent).

The choice of **1**, **2**, **3**, or **4** depends on circumstances. One sensible strategy is to adopt course **3** unless there are particular reasons for doing otherwise. It is important to realise however that when two or more rounded figures are used in a subsequent calculation the rounding errors (which may be insignificant in single numbers) can mount up to such an extent that the result of the calculation may be significantly wrong. The general rule is therefore that figures should be rounded sufficiently to cut out spurious accuracy but no more. It must also be borne in mind that rounding involves an extra process (which must be done *after* checking that the *original* figures sum to the *original* row and column totals) and therefore extra cost. This extra cost may not be justified in all circumstances.

Some possible reasons for dealing differently with the problem of intrinsic errors are as follows:

Reasons   *Course to be adopted*

Insufficient information exists about size of error. (To round the figures one needs to know at least that they are unlikely to be accurate to within so many per cent)

Rounding may be uneconomical or impracticable for some reason

} Course **2** or, occasionally, if all readers are known to be aware that errors exist, course **1**

Errors are known to be negligible   Course **1**

Probable error sizes are known and are sufficiently important to the users of the table to be stated explicitly   Course **4**

The main disadvantage of course **4** is that it can be lead to a complicated presentation, as for example, in the lower half of Table 5.4 from *Family Spending, a report on the 1992 Family Expenditure Survey*. Here shading has been used effectively to connect the entries to their standard errors and the table remains comparatively easy to use.

---

[1] A full definition of these terms can be found in most statistical textbooks but all the non-technical reader of this book needs to know is that a 95 per cent confidence interval is a range which has a 95 per cent probability of including the true value and that the extreme points of such a range are usually about two standard errors above and below the central value.

Table 5.4 Expenditure of all households by tenure of household

| | Rented unfurnished | | | Rented furnished | Rent-free | Owner occupied | | All tenures |
	Local authority	Housing association	Other			In process of purchase	Owned outright	
Total number of households	1,603	238	267	279	149	3,081	1,801	7,418
Total number of persons	3,803	536	580	518	348	8,925	3,464	18,174
Total number of adults	2,617	367	452	426	266	6,208	3,227	13,563
Average number of persons per household								
All persons	2.4	2.3	2.2	1.9	2.3	2.9	1.9	2.5
Males	1.1	1.0	1.1	1.0	1.1	1.4	0.9	1.2
Females	1.3	1.2	1.1	0.9	1.2	1.5	1.0	1.3
Adults	1.6	1.5	1.7	1.5	1.8	2.0	1.8	1.8
Persons under 65	1.2	1.1	1.2	1.5	1.3	2.0	0.9	1.5
Persons 65 and over	0.5	0.4	0.5	-	0.5	0.1	0.8	0.4
Children	0.7	0.7	0.5	0.3	0.6	0.9	0.1	0.6
Children under 2	0.1	0.1	0.1	0.1	-	0.1	-	0.1
Children 2 and under 5	0.1	0.2	0.1	0.1	0.1	0.2	-	0.1
Children 5 and under 18	0.5	0.5	0.3	0.2	0.4	0.6	0.1	0.4
Persons economically active	0.7	0.8	1.0	1.1	1.1	1.7	0.7	1.2
Persons not economically active	1.6	1.5	1.2	0.8	1.3	1.2	1.2	1.3
Men 65 and over, women 60 and over	0.5	0.4	0.5	-	0.4	0.1	0.9	0.4
Others	1.1	1.1	0.7	0.8	0.8	1.1	0.3	0.9
Average age of head of household	52	51	52	30	52	42	66	50
Average weekly household expenditure (£)								
Commodity or service								
Group totals								
Housing : Gross	41.14	48.68	57.26	72.40	14.59	78.58	24.51	54.12
Percentage standard error	*0.8*	*3.2*	*5.7*	*3.4*	*14.5*	*1.6*	*3.0*	*1.2*
Net	22.07	27.34	43.73	55.46	12.60	77.39	23.29	47.36
*	*2.1*	*5.9*	*7.6*	*4.7*	*16.9*	*1.7*	*3.2*	*1.4*
Fuel, light and power	12.00	11.32	11.68	9.77	12.42	14.36	12.62	13.02
*	*1.4*	*3.4*	*4.5*	*5.1*	*6.6*	*1.2*	*1.4*	*0.7*
Food	35.66	37.03	39.85	36.59	46.31	59.26	42.88	47.66
*	*1.6*	*4.4*	*3.9*	*4.5*	*4.9*	*0.9*	*1.5*	*0.7*
Alcoholic drink	7.32	5.24	9.01	12.96	7.37	15.15	8.49	11.06
*	*4.9*	*10.9*	*9.5*	*8.2*	*12.3*	*2.2*	*3.8*	*1.7*
Tobacco	7.94	6.43	5.40	5.15	5.05	5.29	3.17	5.38
*	*3.2*	*9.0*	*11.0*	*9.4*	*14.9*	*3.1*	*5.3*	*2.0*
Clothing and footwear	9.51	9.99	11.28	13.00	12.68	22.89	13.85	16.39
*	*5.1*	*12.3*	*9.1*	*10.5*	*11.6*	*2.5*	*4.7*	*1.9*
Household goods	12.66	12.61	13.82	7.97	15.69	30.27	20.89	21.90
*	*4.2*	*11.7*	*8.9*	*10.1*	*12.1*	*3.2*	*6.2*	*2.5*
Household services	6.03	6.91	10.02	10.46	9.92	18.15	13.92	13.40
*	*5.0*	*8.1*	*9.5*	*9.3*	*12.0*	*2.6*	*5.8*	*2.2*
Personal goods and services	5.52	6.27	7.87	7.56	9.38	13.71	9.60	10.18
*	*4.0*	*8.5*	*8.3*	*9.9*	*16.0*	*2.4*	*3.8*	*1.8*
Motoring expenditure	11.05	14.96	30.44	31.79	33.58	52.99	32.19	35.66
*	*5.5*	*15.4*	*15.0*	*20.9*	*14.9*	*3.3*	*6.7*	*2.8*
Fares and other travel costs	4.44	6.56	8.52	9.83	3.28	10.64	3.57	7.20
*	*6.6*	*33.2*	*34.4*	*13.3*	*22.9*	*20.3*	*11.0*	*12.7*
Leisure goods	7.27	7.11	10.17	11.14	11.24	18.71	11.29	13.32
*	*5.4*	*15.7*	*13.0*	*12.0*	*11.5*	*4.5*	*5.8*	*3.1*
Leisure services	8.99	12.50	17.02	22.22	21.84	34.32	37.38	27.56
*	*5.2*	*15.4*	*13.4*	*16.7*	*16.7*	*3.5*	*31.5*	*10.6*
Miscellaneous	1.06	0.85	1.13	0.71	1.67	2.74	1.05	1.75
*	*23.3*	*25.4*	*20.9*	*28.2*	*29.4*	*5.1*	*13.4*	*5.0*
All expenditure groups	151.53	165.13	219.93	234.61	203.02	375.86	234.19	271.83
Percentage standard error	*1.7*	*4.9*	*5.2*	*5.2*	*5.8*	*1.2*	*5.6*	*1.5*
Average weekly expenditure per person(£)								
All expenditure groups	63.88	73.32	101.26	126.34	85.59	166.90	107.82	146.38

5.7 Footnotes and explanatory notes covering a set of tables

The prime objective of footnotes and explanatory notes is to prevent misuse or misinterpretation of the data. They should therefore be used to record:

- any change in the coverage of entries in the table (for example 'Excludes the Channel Islands and the Isle of Man from 1988')

- any change in the definition of terms used in the table (for example '38 polytechnics and central institutions obtained university status in 1992')

- further explanation of terms used in the table (for example 'Includes GCSE/GCE O Level/SCE grades A - C and CSE grade 1'

- any difference in status of some entries in the table (for example 'Figures for the latest year are provisional')

- explanation of conventions used in arriving at entries (for example 'The seasonally adjusted figures do not always add to the calendar year total which is the sum of unadjusted quarterly figures').

Footnotes to reference tables must obviously be clear and complete. If you are responsible for the design of a set of related tables, the same footnote(s) may apply to a number of them. It is then tempting to introduce a set of tables with the explanatory notes which are common to all or several tables. This practice should be used conservatively. Many readers will refer to a single table in the middle of the set and may not bother to consult the beginning or end of the set of tables for covering notes. If they are in a hurry they may guess at the correct interpretation of an unclear entry rather than search for the appropriate explanatory note. It is therefore preferable to repeat the same footnote(s) after a number of tables rather than to preface a set of tables with general explanatory notes. If such notes are too extensive to be printed underneath a number of tables, then a concise footnote can be included referring users to the appropriate explanatory notes (for example 'See additional Notes 1, 2, and 5 on page x').

Footnotes may be used simply to amplify a row or column heading which was too cumbersome to print in full in the main table. For example, a column headed 'pulp, paper etc' might be used with the accompanying footnote 'pulp, paper and paper products; publishing and printing'. The physical arrangement of footnotes is discussed in Chapter 3, page 26.

5.8 Definitions

A list of definitions should be used to give precise explanations of terms which are used in a set of tables: for example, terms like 'Temporarily stopped workers', 'Weekly hours worked', and 'Short-time working' in tables of employment statistics. This list should also include definitions of any technical terms used in the tables and should normally appear with the entries in alphabetical order immediately after the tables. Figure 5.1 shows the list of definitions printed at the end of the tables in the Employment Department's *Employment Gazette*.

5.9 Treatment of years

It is important to record the exact time period over which data have been collected. The 12-month period beginning 1 January 1993 is clearly labelled as 1993; however data are often recorded for financial years, academic years or other 12-month periods. In such cases, a consistent treatment should be used and should be explained to the reader. A convention, commonly used in the Government Statistical Service and adopted in this handbook, is as follows:

1993-94	(with hyphen) denotes the financial year April 1993 to March 1994
1993/94	(with oblique stroke) denotes the academic year September 1993 to August 1994
1993-1994	(with hyphen and all four digits given for the second year) denotes the 2-year span January 1993 to December 1994.

5.10 Key to abbreviations and special symbols used

When appropriate, a key to abbreviations and symbols should be included either at the beginning or the end of any set of reference tables, preferably on a page by itself for immediate visibility. A typical list might look like this:

..	not available
-	nil or negligible (less than half the final digit shown)
e	estimated
r	revised
n.e.s.	not elsewhere specified
p	provisional
SIC	UK Standard Industrial Classification 1992
EU	European Union

(See for example the panel of conventions in Figure 5.1.)

Obviously it is important to use the same symbols and abbreviations in all tables and, if possible, to choose ones which are compatible with those used in similar tables elsewhere.

5.11 Notes on method of data collection

Reference tables will be used by researchers who may need to know exactly how the data were collected: whether by a complete census or a sample survey; how the sample was selected; how respondents were stratified; what the response rates were in different sub-samples; whether questionnaires were filled in by an interviewer or by the respondent; exactly how questions were worded and what, if any, prompts were used. If the data come from administrative records the researcher may need to know how the records are compiled and exactly how they have been used to produce the tabulated data.

If the table is a one-off production, it is probably best to include such information in a technical appendix. If the table is one of a regularly published series, all the necessary information about the methods used to collect the data should be easily accessible to researchers. This may be in the form of a separate booklet such as, the technical handbook published by HMSO which describes the methods used to collect data in the Family Expenditure Survey[1]. Alternatively a compact description may be included in the covering notes along with an address and telephone number for further enquiries, or an address and telephone number for further enquiries can be given alone so long as up-to-date copies of a suitable document are readily available.

5.12 Commentary on reference tables

Strictly speaking, reference tables do not need a commentary. They provide data, well laid out and clearly explained, for users to extract and analyse according to their individual needs.

Nevertheless there are circumstance in which it is common to publish a commentary along with reference tables. These are:

a   in regular statistical publications, such *Statistical Bulletins*[2] and the *Employment Gazette*[3]. Here tables or sets of tables are usually introduced by a brief commentary on trends and changes revealed in the latest tables

---

[1] KEMSLEY, WFF, REDPATH, RU and HOLMES, M *Family Expenditure Survey Handbook*, Social Survey Division OPCS

[2] *Statistical Bulletins* on a variety of economic statistics are published by the Central Statistical Office (now the Office of National Statistics)

[3] *Employment Gazette* is published monthly by the Employment Department

Figure 5.1 Extract from Employment Gazette (reduced to 90 per cent of original size)

DEFINITIONS

CLAIMANT UNEMPLOYED

People claiming benefit, i.e. Unemployment Benefit, Income Support or National Insurance credits at Unemployment Benefit Offices on the day of the monthly count, who say on that day they are unemployed and that they satisfy the conditions for claiming benefit. (Students claiming benefit during a vacation and who intend to return to full-time education are excluded.)

EARNINGS

Total gross remuneration which employees receive from their employers in the form of money. Income in kind and employers' contributions to National Insurance and pension funds are excluded.

ECONOMICALLY ACTIVE

In *tables 7.1, 7.2* and *7.3* (Labour Force Survey) people aged 16 and over who are in employment (as employees, self employed, on government employment and training programmes, or from 1992, as unpaid family workers) together with those who are ILO unemployed.

ECONOMICALLY INACTIVE

In *tables 7.1, 7.2* and *7.3* (Labour Force Survey) people aged 16 and over who are neither in employment nor ILO unemployed; this group includes people who are, for example, retired or looking after their home/family.

EMPLOYEES IN EMPLOYMENT

A count of civilian jobs of employees paid by employers who run a PAYE scheme. Participants in Government employment and training schemes are included if they have a contract of employment. HM Forces, homeworkers and private domestic servants are excluded. As the estimates of employees in employment are derived from employers' reports of the number of people they employ, individuals holding two jobs with different employers will be counted twice.

FULL-TIME WORKERS

People normally working for more than 30 hours a week except where otherwise stated.

GENERAL INDEX OF RETAIL PRICES

The general index covers almost all goods and services purchased by most households, excluding only those for which the income of the household is in the top 4 per cent and those one and two person pensioner households (covered by separate indices) who depend mainly on state benefits, i.e. more than three-quarters of their income is from state benefits.

HM FORCES

All UK service personnel of HM Regular Forces, wherever serving, including those on release leave.

ILO UNEMPLOYED

In *tables 7.1, 7.2* and *7.3* (Labour Force Survey) people without a paid job in the reference week who were available to start work in the next fortnight and who either looked for work at some time in the last four weeks or were waiting to start a job already obtained.

INDUSTRIAL DISPUTES

Statistics of stoppages of work due to industrial disputes in the United Kingdom relate only to disputes connected with terms and conditions of employment. Stoppages involving fewer than 10 workers or lasting

The terms used in the tables are defined more fully in the periodic articles in Employment Gazette *which relate to particular statistical series.*

less than one day are excluded except where the aggregate of working days lost exceeded 100.

Workers involved and working days lost relate to persons both directly and indirectly involved (thrown out of work although not parties to the disputes) at the establishments where the disputes occurred. People laid off and working days lost elsewhere, owing for example to resulting shortages of supplies, are not included.

There are difficulties in ensuring complete recording of stoppages, in particular those near the margins of the definitions; for example, short disputes lasting only a day or so. Any under-recording would particularly bear on those industries most affected by such stoppages, and would affect the total number of stoppages much more than the number of working days lost.

MANUAL WORKERS (OPERATIVES)

Employees other than those in administrative, professional, technical and clerical occupations.

MANUFACTURING INDUSTRIES

SIC 1980 Divisions 2 to 4.

NORMAL WEEKLY HOURS

The time which the employee is expected to work in a normal week, excluding all overtime and main meal breaks. This may be specified in national collective agreements and statutory wages orders for manual workers.

OVERTIME

Work outside normal hours for which a premium rate is paid.

CONVENTIONS

The following standard symbols are used:

..	not available
–	nil or negligible (less than half the final digit shown)
P	provisional
—	break in series
R	revised
r	series revised from indicated entry onwards
nes	not elsewhere specified
SIC	UK Standard Industrial Classification, 1980 edition
EC	European Community

Where figures have been rounded to the final digit, there may be an apparent slight discrepancy between the sum of the consitilent items and the total as shown. Although figures may be given in unrounded form to facilitate the calculation of percentage changes, rates of change etc by users, this does not imply that the figures can be estimated to this degree of precision, and it must be recognised that they may be the subject of sampling and other errors.

PART-TIME WORKERS

People normally working for not more than 30 hours a week except where otherwise stated.

PRODUCTION INDUSTRIES

SIC 1980 Divisions 1 to 4.

SEASONALLY ADJUSTED

Adjusted for regular seasonal variations.

SELF-EMPLOYED PEOPLE

Those who in their main employment work on their own account, whether or not they have any employees. Second occupations classified as self-employed are not included.

SERVICE INDUSTRIES

SIC 1980 Divisions 6 to 9.

SHORT-TIME WORKING

Arrangements made by an employer for working less than regular hours. Therefore time lost through sickness, holidays, absenteeism and the direct effects of industrial disputes is not counted as short-time.

STANDARD INDUSTRIAL CLASSIFICATION (SIC)

The classification system used to provide a consistent industrial breakdown for UK official statistics. It was revised in 1968 and 1980.

TAX AND PRICE INDEX

Measures the increase in gross taxable income needed to compensate taxpayers for any increase in retail prices, taking account of changes to direct taxes (including employees' National Insurance contributions). Annual and quarterly figures are averages of monthly indices.

TEMPORARILY STOPPED

People who at the date of the unemployment count are suspended by their employers on the understanding that they will shortly resume work and are claiming benefit. These people are not included in the unemployment figures.

VACANCY

A job opportunity notified by an employer to a Jobcentre or Careers Office (including 'self employed' opportunities created by employers) which remained unfilled on the day of the count.

WEEKLY HOURS WORKED

Actual hours worked during the reference week and hours not worked but paid for under guarantee agreements.

WORKFORCE

Workforce in employment plus the claimant unemployed as defined above.

WORKFORCE IN EMPLOYMENT

Employees in employment, self-employed, HM Forces and participants on work-related Government training programmes.

WORK-RELATED GOVERNMENT TRAINING PROGRAMMES

Those participants on Government programmes and schemes who in the course of their participation receive training in the context of a workplace but are not employees, self-employed or HM Forces.

b when a specific investigation has been undertaken in order to provide reference data in a previously unquantified area. Here it is customary to accompany the tables with a summary of the main patterns revealed by the investigation. (For example, a survey conducted by the Office of Population Censuses and Surveys (OPCS) in order to find out how many mothers breastfed their babies and what factors affected the choice of breast or bottle feeding, might be used to provide data for future analyses and comparisons, but the main findings of the survey would certainly be reported in some detail.)

The two circumstances are rather different. In the first it is reasonable to assume that the commentary is intended for regular users of this particular data set and so all that is needed is a brief statement of the latest changes. In the second, the data have been collected in order to chart a previously unquantified area and the report should provide a clear overview of what was discovered as well as more detailed commentary on individual tables.

The first sort of example actually occurs when a single table is used both for demonstration and reference purposes. Although this approach is not recommended when writing for general readers, it may be quite acceptable in specialist publications where most readers are sufficiently familiar with the data to interpret the figures from a reference-type table.

In the second example it is quite possible for the overview report to include additional demonstration tables and charts, based on the reference tables but designed so as to highlight points of particular interest.

The subject of report writing is beyond the scope of this book but some guidelines on writing about numbers are discussed in Chapter 8.

5.13 Some further examples

Finally, it is instructive to examine some tables with a critical eye.

The first example has already appeared as Table 5.1, but it is repeated here as Table 5.5 for ease of reference. This table is 'Gross domestic product by category of expenditure' from the Central Statistical Office's[1] monthly publication *Economic Trends*. It is primarily a reference table, but has been laid out well and, without compromising its main purpose, it also serves to demonstrate effectively the changes in volume of each category of expenditure. The table heading and header detail can be criticised for skimping on such important items as the geographical coverage (is it GB or UK?) but the main body of the table is superb. It illustrates in particular the following points of style:

1 horizontal lines are kept to a functional minimum and there are no vertical lines

2 columns are evenly spaced irrespective of length of column headings. This has been achieved by spreading long words such as 'expenditure' over two lines and long column headings over many lines

3 character size and row and column spacing have been carefully chosen to enable as much information as possible to be clearly set out on the page

4 the percentage changes in the second half of the table being different entities from the values shown in the first part are shown in italics

5 in numbers greater than 1,000 a narrow space has been used, rather than a comma, to separate the thousands from the hundreds.

---

[1] Now the Office for National Statistics

Table 5.5 Gross domestic product: by category of expenditure

£ million, 1990 prices

| | Domestic expenditure on goods and services at market prices | | | | | | | | | | Statistical discrepancy (expenditure adjustment)[4] | Gross domestic product at market prices[1] | less Factor cost adjustment[3] | Gross domestic product at factor cost[4] |
| | Consumers' expenditure+[1] | General government final consumption | | | Gross domestic fixed capital formation+ | Value of physical increase in stocks and work in progress+[2] | Total domestic expenditure[1] | Exports of goods and services+ | Total final expenditure[1] | less Imports of goods and services+ | | | | |
		Central government	Local authorities	Total										
	CCBH	DJDK	DJDL	DJCZ	DFDM	DHBK	DIEL	DJCV	DJDA	DJCY	GIXS	CAOO	DJCU	CAOP
1989	345 406	68 836	41 303	110 139	110 503	3 669	569 717	126 836	696 553	147 615	–	548 938	72 712	476 226
1990	347 527	70 108	42 826	112 934	106 776	−1 118	566 119	133 284	699 403	148 285	–	551 118	72 232	478 886
1991	339 993	71 950	43 847	115 797	96 265	−4 722	547 333	132 114	679 447	140 248	−430	538 769	71 049	467 720
1992	339 610	72 678	43 957	116 635	94 741	−1 773	549 213	135 547	684 760	148 271	−668	535 821	70 250	465 571
1993	..	..	..	..	..	..	..	..	..	..	..	..	..	474 665
	CAAB	DIAV	DIAW	DIAT	DECU	DGBA	DIAY	DJDG	DIAU	DJDJ			DIAS	
1988 Q3	84 690	16 592	10 276	26 868	26 288	1 322	139 168	30 707	169 875	34 719	–	135 156	17 943	117 213
Q4	85 562	17 046	10 215	27 261	26 629	3 469	142 921	30 032	172 953	36 602	–	136 351	18 248	118 103
1989 Q1	85 847	17 023	10 259	27 282	28 343	953	142 425	31 496	173 921	37 133	–	136 788	18 120	118 668
Q2	86 472	16 925	10 260	27 185	27 551	1 485	142 693	31 000	173 693	36 468	–	137 225	18 327	118 898
Q3	86 243	17 545	10 378	27 923	27 449	1 345	142 960	31 759	174 719	37 350	–	137 369	18 239	119 130
Q4	86 844	17 343	10 406	27 749	27 160	−114	141 639	32 581	174 220	36 664	–	137 556	18 026	119 530
1990 Q1	86 992	17 547	10 443	27 990	27 628	27	142 637	33 259	175 896	37 630	–	138 266	18 112	120 154
Q2	87 409	17 484	10 673	28 157	27 124	547	143 237	33 264	176 501	37 487	–	139 014	18 433	120 581
Q3	86 778	17 448	10 824	28 272	26 397	−133	141 314	33 110	174 424	36 881	–	137 543	18 029	119 514
Q4	86 348	17 629	10 886	28 515	25 627	−1 559	138 931	33 651	172 582	36 287	–	136 295	17 658	118 637
1991 Q1	85 834	17 902	10 917	28 819	24 669	−1 078	138 244	31 932	170 176	34 908	−84	135 184	17 707	117 477
Q2	84 806	18 086	10 944	29 030	24 063	−1 692	136 207	33 159	169 366	34 726	−104	134 536	17 657	116 879
Q3	84 712	18 039	10 990	29 029	23 750	−1 535	135 956	33 475	169 431	34 923	−117	134 391	17 715	116 676
Q4	84 641	17 923	10 996	28 919	23 783	−417	136 926	33 548	170 474	35 691	−125	134 658	17 970	116 688
1992 Q1	84 227	18 223	10 983	29 206	23 817	−858	136 392	33 460	169 852	36 273	−145	133 434	17 443	115 991
Q2	84 717	18 525	10 956	29 481	23 548	−733	137 013	34 078	171 091	37 262	−160	133 669	17 589	116 080
Q3	85 089	17 905	11 002	28 907	23 450	274	137 720	33 819	171 539	37 222	−175	134 142	17 579	116 563
Q4	85 577	18 025	11 016	29 041	23 926	−456	138 088	34 190	172 278	37 514	−188	134 576	17 639	116 937
1993 Q1	85 981	17 871	11 048	28 919	24 032	−766	138 166	35 082	173 248	37 766	−155	135 327	17 735	117 592
Q2	86 432	18 765	10 406	29 171	23 386	241	139 230	34 597	173 827	37 424	−153	136 250	18 015	118 235
Q3	87 268	18 699	10 444	29 143	23 560	−360	139 602	35 591	175 193	37 900	−151	137 142	18 140	119 002
Q4	..	..	..	..	..	..	..	..	..	..	..	..	..	119 836

Percentage change, quarter on corresponding quarter of previous year

1988 Q3	7.7	−1.7	0.4	−0.9	10.4		6.9	−0.5	5.5	8.8		4.7	5.0	4.6
Q4	7.1	1.7	−0.2	1.0	7.0		8.5	1.8	7.3	19.2		4.5	5.3	4.4
1989 Q1	5.0	−0.2	−0.3	−0.2	10.7		5.5	5.4	5.5	14.7		3.2	3.1	3.3
Q2	4.7	0.2	0.2	0.2	5.1		4.6	1.4	4.0	8.0		3.0	3.6	2.9
Q3	1.8	5.7	1.0	3.9	4.4		2.7	3.4	2.9	7.6		1.6	1.6	1.6
Q4	1.5	1.7	1.9	1.8	2.0		−0.9	8.5	0.7	0.2		0.9	−1.2	1.2
1990 Q1	1.3	3.1	1.8	2.6	−2.5		0.1	5.6	1.1	1.3		1.1	0.0	1.3
Q2	1.1	3.3	4.0	3.6	−1.5		0.4	7.3	1.6	2.8		1.3	0.6	1.4
Q3	0.6	−0.6	4.3	1.2	−3.8		−1.2	4.3	−0.2	−1.3		0.1	−1.2	0.3
Q4	−0.6	1.6	4.6	2.8	−5.6		−1.9	3.3	−0.9	−1.0		−0.9	−2.0	−0.7
1991 Q1	−1.3	2.0	4.5	3.0	−10.7		−3.1	−4.0	−3.3	−7.2		−2.2	−2.2	−2.2
Q2	−3.0	3.4	2.5	3.1	−11.3		−4.9	−0.3	−4.0	−7.4		−3.2	−4.2	−3.1
Q3	−2.4	3.4	1.5	2.7	−10.0		−3.8	1.1	−2.9	−5.3		−2.3	−1.7	−2.4
Q4	−2.0	1.7	1.0	1.4	−7.2		−1.4	−0.3	−1.2	−1.6		−1.2	1.8	−1.6
1992 Q1	−1.9	1.8	0.6	1.3	−3.5		−1.3	4.8	−0.2	3.9		−1.3	−1.5	−1.3
Q2	−0.1	2.4	0.1	1.6	−2.1		0.6	2.8	1.0	7.3		−0.6	−0.4	−0.7
Q3	0.4	−0.7	0.1	−0.4	−1.3		1.3	1.0	1.2	6.6		−0.2	−0.8	−0.1
Q4	1.1	0.6	0.2	0.4	0.6		0.8	1.9	1.1	5.1		−0.1	−1.8	0.2
1993 Q1	2.1	−1.9	0.6	−1.0	0.9		1.3	4.8	2.0	4.1		1.4	1.7	1.4
Q2	2.0	1.3	−5.0	−1.1	−0.7		1.6	1.5	1.6	0.4		1.9	2.4	1.9
Q3	2.6	4.4	−5.1	0.8	0.5		1.4	5.2	2.1	1.8		2.2	3.2	2.1
Q4	..	..	..	..	..		..	..	..	..		..	..	2.5

1 These series are affected by the abolition of domestic rates and the introduction of the Community Charge. For details, see notes in the UK National Accounts article in the latest edition of *UK Economic Accounts*.

2 Includes quarterly alignment adjustment. For explanation see notes in the UK National Accounts article in the latest edition of *UK Economic Accounts*.

3 Represents taxes on expenditure less subsidies, both valued at 1990 prices.

4 GDP is estimated in seasonally adjusted form only. Therefore whilst seasonally and unadjusted versions exist of the residual error, the attribution of statistical discrepancies to the expenditure-based and income-based estimates can be made only in seasonally adjusted form.

Source: Central Statistical Office

The remaining examples all illustrate undesirable features. In Table 5.6, the print is so small as to be illegible (though later editions of the *UK Economic Accounts* enlarged the table). In Table 5.7, the table has been drawn to fill the full width of the page regardless of the convenience of the reader. However, it helpfully gives a telephone number for enquiries. Table 5.8 has too many extraneous lines (see also Table 3.12) and, although Table 5.9 looks professional, it is desperately in need of firebreaks and a smaller gap between the 'Year' column and the data columns. The table would also be improved to a smaller extent if the commas separating thousands and hundreds were replaced with spaces and the percentages were shown in italics or separated slightly from other columns. The geographical coverage is not explicit in the table, but the notes at the beginning of the publication explain that all tables are for the UK unless stated otherwise. This saves repetition within the publication, but will not help the reader who receives a photocopy of a single table.

Table 5.6 National Accounts Aggregates

Index numbers 1985=100, seasonally adjusted[1]

	Value indices at current prices		Volume indices at 1985 prices			Implied gross domestic product deflator[4]	
	Gross domestic product at market prices[2]	Gross domestic product at factor cost	Gross national disposable income at market prices[3]	Gross domestic product at market prices	Gross domestic product at factor cost	At market prices	At factor cost[5]
	DJCL	CAON	DJCR	FNAO	DJDD	DJDT	DJCM
1985	100.0	100.0	100.0	100.0	100.0	100.0	100.0
1986	107.7	106.6	103.9	104.1	103.8	103.5	102.6
1987	118.5	117.1	108.3	109.1	108.6	106.6	107.8
1988	131.9	130.3	113.7	114.0	113.5	115.8	114.8
1989	144.2	143.3	116.3	116.4	115.8	124.0	123.7
1990	154.0	155.7	117.1	117.0	116.6	131.7	133.6
1991	160.5	161.2	116.0	114.4	113.7	140.3	141.8
1992	166.5	167.3	115.2	113.7	113.1	146.5	147.9
1989 Q1	140.7	139.8	116.5	115.6	115.4	121.7	121.1
Q2	142.9	141.7	116.7	116.1	115.5	123.1	122.7
Q3	145.5	144.7	116.1	116.7	116.1	124.7	124.7
Q4	147.9	146.9	115.9	117.1	116.5	126.3	126.1
1990 Q1	151.7	151.0	116.5	117.6	117.2	129.0	128.9
Q2	153.3	155.3	117.2	118.2	117.6	129.7	132.0
Q3	155.5	157.9	118.4	116.6	116.2	133.4	135.9
Q4	155.6	158.6	116.2	115.5	115.3	134.8	137.6
1991 Q1	156.9	159.5	115.9	114.9	114.4	136.8	139.4
Q2	159.6	159.6	116.7	114.1	113.4	139.9	140.7
Q3	161.9	162.0	115.9	114.4	113.6	141.6	142.6
Q4	163.6	163.6	115.7	114.2	113.3	143.3	144.4
1992 Q1	164.4	165.1	115.3	113.5	112.9	144.8	146.2
Q2	166.5	167.4	114.9	113.4	112.8	146.9	148.3
Q3	167.8	168.8	116.1	113.8	113.2	147.5	149.1
Q4	167.4	168.0	114.4	114.2	113.5	146.6	148.0
1993 Q1	169.4	169.9	114.6	114.7	114.0	147.7	149.1

Percentage change, latest quarter on previous quarter

1993 Q1	1.2	1.1	0.1	0.5	0.4	0.7	0.7

Percentage change, latest quarter on corresponding quarter of previous year year

1993 Q1	3.0	2.9	−0.6	1.1	0.9	2.0	1.9

1 These estimates are given to one decimal place but this does not imply that they can be regarded as accurate to the last digit shown. Estimates at current market prices are affected by the abolition of domestic rates and the introduction of the community charge.
2 "Money GDP".
3 Also known as Real national disposable income (RNDI).
4 Based on sum of expenditure components of GDP at current and constant prices.
5 Also known as the Index of total home costs.

Table 5.7 Driving tests: distribution of tests by age, sex and class of vehicle[1]

Percentage

By age:	Motor car		Two wheeler	
	October 1980	April 1985	October 1980	April 1985
Under 21				
Male	27	28	62	49
Female	21	23	4	5
21 to 40				
Male	16	16	26	32
Female	26	26	3	5
41 to 50				
Male	2	}	3	}
Female	4	} 5[2]	1	} 5[3]
Over 50				
Male	1	}	1	}
Female	3	} 3[2]		} 4[3]
All ages				
Male	46	46	92	87
Female	54	54	8	13
All candidates	100	100	100	100

1 Figures are based on all tests conducted during a 2 day period in the months shown.
2 Of candidates over 40, 2 per cent were male and 6 per cent female.
3 Of candidates over 40, 6 per cent were male and 3 per cent female.

Contact point for further
information: 01-276 8208

Table 5.8   Consumers' expenditure on energy
United Kingdom

£ million

	1988	1989	1990	1991	1992
At current market prices					
Coal(2)	715r	655r	584r	633r	660
Coke	91	81	72	74r	46
Gas	4,562r	4,454r	4,892r	5,777r	5,684
Electricity	5,412r	5,878r	6,280r	7,260r	7,670
Petroleum products(3)	335r	391r	493r	528r	499
All fuel and power(4)	11,115r	11,459r	12,321r	14,272r	14,559
Motor spirit, derv fuel and lubricating oil	8228	9,059	10,172r	10,793r	11,044
Total energy products(5)	**19,343r**	**20,518r**	**22,493r**	**25,065r**	**25,603**
Total consumers' expenditure	**302,057r**	**330,532r**	**350,411r**	**367,527r**	**386,606**
Revalued at 1985 prices					
Coal(2)	765r	688r	596r	633r	587
Coke	88	77	70r	65r	38
Gas	4,499r	4,195r	4,304r	4,762r	4,693
Electricity	5,047r	5,106r	5,052r	5,316r	5,326
Petroleum products(3)	469r	498r	531r	602r	648
All fuel and power(4)	10,868r	10,564r	10,573r	11,378r	11,292
Motor spirit, derv fuel and lubricating oil	9,484	9,793	10,043	9,953	9,943
Total energy products(5)	**20,352r**	**20,357r**	**20,616r**	**21,331r**	**21,235**
Total consumers' expenditure	**264,096r**	**272,917r**	**274,744r**	**269,083r**	**269,575**

(1) These figures are based on "National Income and Expenditure in the Fourth Quarter of 1992" an article in the April 1993 issue of Economic Trends published by the Central Statistical Office. The figures exclude business expenditure.
(2) Including some manufactured fuel.
(3) Excluding motor spirit, derv fuel and lubricating oil.
(4) Including an estimate for wood.
(5) Quarterly data on energy expenditure are shown in the Central Statistical Office's Monthly Digest of Statistics.

Table 5.9 Inland Revenue contribution to central government tax revenue
Payments into the consolidated fund

Amounts: £ million

Year	Inland Revenue	Customs & Excise	Vehicle excise duties	Selective employment tax and NI surcharge	Total[1]	Inland Revenue as % of total
1908-09	96[2]	29	-	-	125	76.7
1918-19	622	162	-	-	784	79.3
1928-29	407	253	4	-	664	61.3
1938-39	520	341	36	-	896	58.0
1948-49	2,058	1,557	53	-	3,668	56.1
1958-59	3,016	2,191	107	-	5,314	56.8
1968-69	6,532	4,601	393	1,363	12,888	50.7
1969-70	7,476	4,953	417	1,888	14,733	50.7
1970-71	8,175	4,709	421	1,990	15,295	53.4
1971-72	9,134	5,325	473	1,324	16,256	56.2
1972-73	9,245	5,744	485	994	16,467	56.1
1973-74	10,633	6,220	534	45	17,431	61.0
1974-75	14,191	7,407	532	2	22,132	64.1
1975-76	18,159	9,176	781	-	28,117	64.6
1976-77	20,710	10,900	846	-	32,456	63.8
1977-78	21,917	12,284	1,072	1,163	36,436	60.2
1978-79	24,080	13,835	1,113	1,914	40,943	58.8
1979-80	28,153	18,032	1,149	2,987	50,321	55.9
1980-81	32,982	22,095	1,419	3,542	60,038	54.9
1981-82	40,318	25,428	1,640	3,596	70,983	56.8
1982-83	43,794	27,896	1,823	2,831	76,344	57.4
1983-84	45,926	31,434	2,036	1,671	81,067	56.7
1984-85	50,350	35,502	2,219	924	88,996	56.6
1985-86	55,438	37,398	2,432	35	95,303	58.2
1986-87	57,116	41,094	2,575	1	100,786	56.7
1987-88	64,509	44,738	2,645	-	111,892	57.7
1988-89	68,813	49,565	2,811	2	121,191	56.8
1989-90	76,674	52,190	2,918	1	131,783	58.2
1990-91	82,322	55,337	2,972	1	140,632	58.5
1991-92	79,510	61,827	2,946	-	144,283	55.1
1992-93	76,346	63,398	3,196	-	142,940	53.4
1993-94[3]	76,700	67,500	3,700	-	147,900	51.9

[1] The total is calculated as the sum of the duties shown in this table and excludes some smaller sources of revenue.

[2] Includes £34 million in respect of excise duties which were administered by Customs & Excise in later years.

[3] 1993-94 Financial Statement and Budget Report forecast.

Table 5.10 is a perfectly satisfactory reference table from *Population Trends*, but even here the layout could be improved from a user's point of view by transposing the main row and column variables. At present the table has age as the main column variable with marital status as the sub-category making 24 columns in all. As 24 columns do not fit across the page, the table has been split in half after column 12 and the second half placed underneath the first. This makes it difficult to look at patterns for males and females across all age groups. If, instead, sex became the main column heading with marital status as the sub-category, there would only be eight columns. Age would become the main row variable with time as the sub-category. The revised table would be longer and narrower, which might not suit the house style: it might also tempt the printer to spread the eight columns across the full width of the page. In fact it would be helpful to use the space to add three extra columns of totals for marital status: males, females and total population. At present the reader needs to refer back two pages for these (and do some addition). Other improvements would be to place 'all ages' at the end of the table (where people expect to find totals) and to use spaces rather than commas to separate thousands from hundreds. The effect of these improvements can be seen in Table 5.11.

Table 5.10 Population: by age, sex and marital status

thousands

Mid-year	All ages 16 and over				16–24				25–34			
	Single	Married	Divorced	Widowed	Single	Married	Divorced	Widowed	Single	Married	Divorced	Widowed
Males												
1971	4,173	12,522	187	682	2,539	724	3	–	637	2,450	38	4
1981	5,013	12,238	611	698	3,095	485	10	1	906	2,508	151	4
1991	6,705	11,718	1,221	685	3,139	254	12	–	1,765	2,053	245	2
1992	5,993	11,830	1,236	694	3,029	249	12	–	1,784	2,099	252	3
Females												
1971	3,583	12,566	296	2,810	1,907	1,255	9	2	326	2,635	63	12
1981	4,114	12,284	828	2,939	2,530	904	27	2	496	2,791	218	13
1991	4,808	11,867	1,497	2,925	2,668	542	30	1	1,151	2,472	312	8
1992	4,724	11,941	1,511	2,940	2,569	529	29	1	1,159	2,515	318	8

Mid-year	35–44				45–64				65 and over			
	Single	Married	Divorced	Widowed	Single	Married	Divorced	Widowed	Single	Married	Divorced	Widowed
Males												
1971	317	2,513	48	13	502	4,995	81	173	179	1,840	17	492
1981	316	2,519	178	12	480	4,560	218	147	216	2,167	54	534
1991	473	2,667	388	12	454	4,387	471	122	244	2,358	106	549
1992	468	2,613	380	11	466	4,496	486	122	246	2,373	107	558
Females												
1971	201	2,529	66	48	569	4,709	125	733	580	1,437	32	2,016
1981	170	2,540	222	41	386	4,358	271	620	533	1,692	90	2,263
1991	274	2,766	444	34	286	4,221	538	483	430	1,866	174	2,400
1992	271	2,712	436	33	292	4,326	555	484	432	1,858	174	2,415

Population estimates by marital status for 1991 and 1992 are provisional. For each age by sex group, marital status proportions were calculated from the mid-1991 marital status estimates rolled-forward from the 1981 Census. These proportions were applied to corresponding age by sex groups (all statuses combined) from the mid-1991 and mid-1992 rebased estimates. Estimates derived from statements about marital status made in the 1991 Census of Population, supplemented by information from other sources, will be prepared as soon as possible.

Table 5.11 Population: by age, sex and marital status (redesigned)

England and Wales — thousands

Mid-year Aged	Males					Females					Total Population
	Single	Married	Divorced	Widowed	Total	Single	Married	Divorced	Widowed	Total	
16-24											
1971	2 539	724	3	–	3 268	1 907	1 255	9	2	3 173	6 441
1981	3 095	485	10	1	3 591	2 530	904	27	2	3 163	6 754
1991	3 139	254	12	–	3 405	2 668	542	30	1	3 241	6 646
1992	3 029	249	12	–	3 290	2 569	529	29	1	3 128	6 418
25-34											
1971	637	2 450	38	4	3 129	326	2 635	63	12	3 036	6 164
1981	906	2 508	151	4	3 569	496	2 791	218	13	3 517	7 086
1991	1 765	2 053	245	2	4 065	1 151	2 472	312	8	3 943	8 008
1992	1 784	2 099	252	3	4 138	1 159	2 515	318	8	4 001	8 137
35-44											
1971	317	2 513	48	13	2 891	201	2 529	66	48	2 845	5 736
1981	316	2 519	178	12	3 024	170	2 540	222	41	2 972	5 996
1991	473	2 667	388	12	3 539	274	2 766	444	34	3 517	7 056
1992	68	2 613	380	11	3 472	271	2 712	436	33	3 452	6 924
45-64											
1971	502	4 995	81	173	5 751	569	4 709	125	733	6 136	11 887
1981	480	4 560	218	147	5 405	386	4 358	271	620	5 635	11 040
1991	454	4 387	471	122	5 434	286	4 221	538	483	5 528	10 962
1992	466	4 496	486	122	5 570	292	4 326	555	484	5 657	11 227
65 and over											
1971	179	1 840	17	492	2 528	580	1 437	32	2 016	4 065	6 593
1981	216	2 167	54	534	2 971	533	1 692	90	2 263	4 578	7 549
1991	244	2 358	106	549	3 257	430	1 866	174	2 400	4 870	8 127
1992	246	2 373	107	558	3 274	432	1 858	174	2 415	4 879	8 153
All ages 16 and over											
1971	4 173	12 522	187	682	17 564	3 583	12 566	296	2 810	19 255	36 819
1981	5 013	12 238	611	698	18 560	4 114	12 284	828	2 939	20 165	38 725
1991	6 705	11 718	1 221	685	20 329	4 808	11 867	1 497	2 925	21 097	41 426
1992	5 993	11 830	1 236	694	19 753	4 724	11 941	1 511	2 940	21 116	40 869

Population estimates by marital status for 1991 and 1992 are provisional. For each age by sex group, marital status proportions were calculated from the mid-1991 marital status estimates rolled-forward from the 1981 Census. These proportions were applied to corresponding age by sex groups (all statuses combined) from the mid-1991 and mid-1992 rebased estimates. Estimates derived from statements about marital status made in the 1991 Census of Population, supplemented by information from other sources, will be prepared as soon as possible.

6 Charts

6.1 Do charts work?

Charts are popular with both authors and readers of reports and technical articles. But they often fail as tools of communication. It is useful to probe this paradox by discussing the reasons for both their popularity and their frequent failure.

First, why are they popular with those who construct them? There seem to be two reasons. One reason is that the originator of a chart usually expects it to please the reader. The other, and more important, reason is that once a set of data has been studied and the main pattern in the data clearly identified, a well planned chart (or set of charts) is often the best way of describing these patterns concisely and memorably. Sometimes a picture may indeed be worth a thousand words.

Does it please the reader? Pause for a moment and consider your own reaction on turning a page and finding a clear, apparently simple graph on the next page. Many people will experience two reactions: first relief that there is less than a full page of text to be read and secondly relief that some part of the argument has been condensed into a simple picture which can be scanned quickly. So a chart *is* likely to please the reader. But is it likely to do its job of communicating effectively? Here the answer is much more doubtful.

Remember the communicator's state of mind when he or she produced the chart? He or she had studied and assimilated the data and identified the main patterns. He or she was, in fact, thoroughly familiar with:

- what the data represented
- the units of measurement
- the geographical coverage
- the time period covered
- the range of variation in the data.

All these facts were already part of his or her store of knowledge and the chart represented, *for the communicator*, a concise and efficient way of arranging the data in order to highlight the salient features against this factual background.

Consider, for example, the chart in Figure 6.1 from page 3 of the *Labour Force Survey Quarterly Bulletin No 4*.

The originator of this chart was completely familiar with the definition of economic activity rates for which the reader is invited to consult the technical note on page 22 of the Bulletin[1]. This reads:

> **Economic activity rate**: the percentage of people aged 16 and over who are economically active.

This is concise but unhelpful in that the reader then has to read the definition of 'economically active' and probably also the definitions of 'in employment' and 'economically inactive'. The last definition is at some distance from the others (on the next page in fact). These definitions are reproduced here as part of Figure 6.1 for convenience.

The originator had also decided to show graphically the percentage change in economic activity rates between surveys from 1984/85 to spring 1993 for men and women separately. He or she clearly knew that the changes in the earlier years were mainly positive and in the later years mainly negative and that the patterns for men and women differed.

In fact the originator clearly knew a great deal more about these data including the original activity rates from which the changes were calculated. Activity rates for the latest two periods are given in the table just above the chart and the reader is also referred to Table 1 on another page. However that table gives the *number* of people who are economically active, but no activity rates or changes in activity rates, so that the reader cannot check his or her understanding.

The person who planned the chart has devised an effective way of highlighting the increases in women's activity rates between 1984/85 and 1988/89, when men's activity rates were little changed. The increase in the activity rate for women between winter 1992/93 and spring 1993 stands out less well.

---

[1] This describes the survey, gives definitions of the terms used in this bulletin and gives further details of the differences between LFS estimates and the equivalents from other sources.

Figure 6.1 Extracts from the Labour Force Survey Quarterly Bulletin

Economic activity

Among people aged 16 and over, 73.2 per cent of men and 52.8 per cent of women (*seasonally adjusted*) were economically active in Winter 1992/93. The chart below shows that between 1984 and 1990, economic activity rates for women increased, while those for men showed little change. Since Spring 1991, activity rates have been decreasing; there have been falls since Autumn 1992 of 0.3 percentage points in the male economic activity rate and no change in the rate for women.

Economic Activity - levels & rates		
(seasonally adjusted)		
	Autumn 1992	Winter 1992/93
Economically active (millions)		
All persons	27.69	27.63
Men	15.64	15.58
Women	12.05	12.05
Economic activity rates (per cent)		
All persons	62.8%	62.7%
Men	73.5%	73.2%
Women	52.8%	52.8%

Concepts and definitions

Economically active: People aged 16 and over who are either in employment or unemployed

Economic activity rate: the percentage of people aged 16 and over who are economically active.

Economically inactive: people who are neither in employment nor unemployed on the ILO measure. This group includes, for example, all those who were looking after a home or retired. Although no estimates appear in this bulletin, for other LFS analyses, this group would also include all people aged under 16.

In employment: people aged over 16 who did some paid work in the reference week (whether as an employee or self-employed); those who had a job that they were temporarily away from (on holiday, for example); those on government employment and training programmes; and those doing unpaid family work.

Note: The technical note on page x describes the survey, gives definitions of the terms used in this bulletin and gives further details of the differences between LFS estimates and the equivalents from other sources.

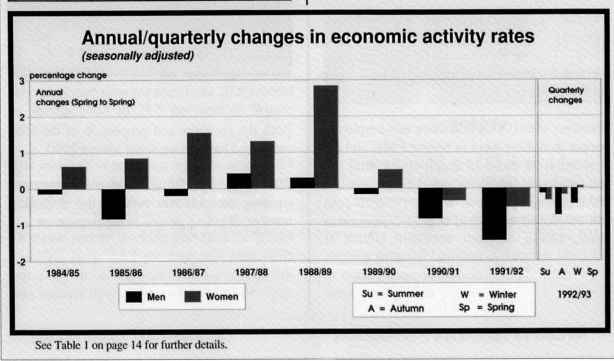

Annual/quarterly changes in economic activity rates
(seasonally adjusted)

See Table 1 on page 14 for further details.

But what about the reader? How long has it taken you to peruse Figure 6.1 and try to match verbal summary to the chart? And are you sure that you understand how the two fit together and exactly which periods are being compared? Without looking back at the figure or the table, can you describe the main pattern in the data using your own words? And can you say whether or not activity rates for men decreased in the last three periods, or whether the activity rates for men and women were greater in spring 1993 than in Spring 1986?

To retrieve this information from the graph the reader would have to weigh up mentally the relative sizes of the increases and decreases. The reader may also be in some doubt about whether the percentage changes in the percentage activity rates are absolute differences or genuine percentage changes, that is whether the difference between 52.8 per cent and 53.1 per cent in the latest period for women is shown as 0.3 per cent (by subtraction) or 0.6 per cent (percentage change). Close examination of the chart is not conclusive, but suggests it is probably the latter.

This is not meant to suggest that the originator of the graph intended the reader to use it to answer such a detailed question. The question is merely posed to emphasise the typical difference between the state of knowledge of those who plan charts and those who read them.

What has just been said illustrates the main limitation on the use of graphs. A graph will communicate its message effectively only to someone who has mastered the necessary background information about units of measurement, the scale used and range covered on each axis as well as any conventions used on the graph (for example, the use of a dotted line to represent, say, the UK figures and a solid line to represent the corresponding figure for a specific region). If you are confident that your reader is familiar with all these details you may then safely rely on a well executed graph to communicate a simple message; you may even choose to use quite a complicated form of graph *so long as both you and your reader are skilled in interpreting its conventions*. But beware of underestimating how long it takes to build up such skill and, if in doubt, assume that few of your audience will be familiar with either the background to the data or any but the most common graphical conventions.

A second problem with using graphs is that most people have not been trained in how to *read* them

rather than merely look at them. A graph looks like a picture and it is well known that we have only to glance at a picture for a second or two in order to recognise it when we meet it again. This encourages us to register only the obvious pattern in a graph before passing back to the text.

On courses at the Civil Service College, the following graph has been shown on the overhead projector:

Figure 6.2   Peak age of known offending, 1991

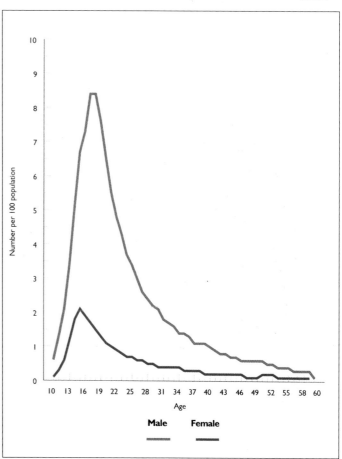

Source: Criminal Statistics, England and Wales, HMSO (reduced in size)

In the following discussion, all the audience remembered the shape and relative position of lines well enough to trace them in the air. But they remembered the peak age of known offending for men as 18 only because it had been stressed by the presenter. They had no memory of the units.

So where does this leave us? Can charts ever be relied on as effective communication tools for a non-specialist audience? Yes indeed; they can work very well provided you recognise the limitations on the sort of message which can be transmitted graphically and give due thought to the choice of chart and how it should be presented.

Charts can work extremely well for showing relationships in the data - features like:

- the comparative sizes of the parts which make up a total

- a dramatic change in one contributor's share of a total

- the comparative sizes of a number of related measures

- a striking pattern of growth or decline over time (in one or possibly several related series)

- a change in the ranked order of a set of related measurements

- a correlation between two sets of data.

Patterns like these can be effectively encapsulated in a well chosen chart, but detailed numerical information cannot. For example, we see immediately from Figure 6.3 that three quarters of freight is carried by road and that there was a decrease in the proportion of goods carried by rail between 1955 and 1990 while the other proportions (road, water, pipeline) increased. These are the messages which the eye registers, not the numbers 75 per cent, 81 per cent, 21 per cent and 6 per cent (even although these figures appear on the chart).

If it is important that the reader remembers these numbers, the chart alone is unlikely to be enough. In fact, this pie chart is unlikely to serve any useful purpose in a report unless it is accompanied by a verbal summary - something like:

The total amount of freight transported in Great Britain increased by over 60 per cent between 1955 and 1990 to 2,160 million tonnes. Over three quarters of these goods were transported by road. Since 1955, the proportion of goods transported by rail has fallen from 21 per cent (in 1955) to 6 per cent in 1990; the proportions transported by road, water and pipeline all increased.

Figure 6.3 Freight transport by mode

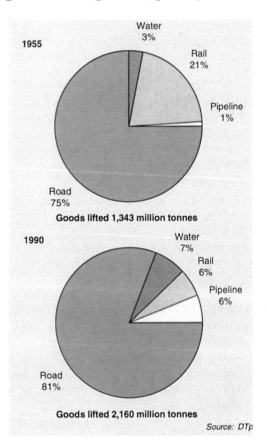

Quoting the actual figures in the summary performs two useful functions. Firstly it confirms to the reader that he or she has interpreted the chart correctly and secondly it links the numbers with the corresponding pie segments in the reader's mind. There is now a fair chance that the whole message will be stored in his or her memory.

This example illustrates the two basic rules for producing an effective chart:

- keep it simple
- talk the reader through it.

6.2 When to use a chart

Good graphs are time consuming and relatively expensive to produce; they should therefore be used only when they offer real benefits in getting the message over. It has been said that graphs should be used to:

- reveal the unexpected
- make the complex easier to perceive.

To this should be added, for the purpose of communicating to a non-specialist audience:

- to convey concisely and memorably a message (which may or may not be unexpected).

Just as in the case of a table, the data must be studied and the salient features decided upon before any attempt is made to represent them graphically. The producer of a chart must have decided in advance what message the chart is to transmit. A good way of doing this is to start by writing a verbal summary of the data: three or four sentences highlighting what you judge to be the main patterns and exceptions will be sufficient. This process is inevitably subjective, but you as a communicator must face up to this. *Any attempt to use a chart as a neutral data store is a waste of time and effort.* You have promised your reader that some data are relevant at this point in your report and so you must demonstrate this by saying clearly 'here is the message in the data'.

There are surprisingly few different sorts of message. All are based on comparisons and most applications can be classified as one of the following types:

Table 6.1 Comparisons and typical summaries

Sort of comparison	Typical summary
Components as part of a whole	In 1991, 20 per cent of the UK population was aged under 16, 64 per cent were aged 16-64 while 16 per cent were 65 or over.
Change in composition of a whole	By 2031, the percentage of the UK population aged under 16 is projected to decline to 18 per cent (from 20 per cent in 1991). Over the same period, the proportion aged 16-64 is also projected to fall, from 64 per cent to 59 per cent, and the proportion aged 65 or over is expected to grow from 16 per cent to 23 per cent.
Comparative sizes of related measurements	In 1991, 19 per cent of the white population in Great Britain was aged under 16 compared with 33 per cent of the ethnic minority population.
Change over time of one or more related measurements	Between 1976 and 1993 the number of male employees fell from 13.4 million to 10.6 million but the number of female employees rose from 9.2 million to 10.2 million.
Change in the ranked order of a set of related measurements	In 1981, the most popular tourist attractions in Great Britain were Blackpool Pleasure Beach (7.5 million admissions), the Science Museum (3.8 million) and the Natural History Museum (3.7 million). In 1991 the Blackpool Pleasure Beach was still the most popular (6.5 million admissions) followed by the British Museum (6.3 million). The Science Museum and Natural History Museum, which now charge admission, had dropped to 11th and 17th place respectively.
Relationship between two sets of data	EU countries with high economic activity rates for men also tend to have high economic activity rates for women. However there is a much greater variation in women's activity rates. The highest overall rate is 68 per cent in Denmark (75 per cent for males and 62 per cent for females); the lowest overall rate is 49 per cent in Greece (66 per cent and 36 per cent for males and females respectively).

All these sorts of comparisons may be made graphically. But whether or not a graph is the most effective way of communicating such a message depends on a number of factors including:

- whether it is more important to leave the reader with a clear impression of a relationship (for example, 'A's share was much bigger that B's but is now smaller') or a detailed numerical message (for example, 'dividends increased from 4p to 7p per share' or 'education accounted for 72 per cent of total expenditure'): generally speaking, graphs convey relationships better than tables or text;

- whether or not the message to be transmitted lends itself to a clear graphical statement (for example, if two series have grown in parallel over a 20-year period this can be shown concisely using a set of line graphs; conversely, four series having dissimilar patterns of growth or decline over a five-year period are likely to produce a confused tangle of lines);

- the scales of measurement (for example, can the range of both variables be shown comfortably on the same scale?);

- the skill with which the chart is planned and executed (skill in planning comes with practice and a continual critical appraisal of all charts, your own and other people's: skill in execution often requires professional help);

- the reader: there is not much you can do about this unless, of course, the report is aimed at a small specific audience in which case it pays to get to know their idiosyncrasies; some people like graphs and others do not, but for both sorts of reader a good chart is more likely to prove successful than a poor one.

6.3 What sort of chart?

Assuming that you have decided to use a chart to illustrate a particular point, what sort of chart should be used? For a non-specialist audience it is probably safest to choose from a limited range of three or four basic varieties:

- pie charts
- bar charts (in a wide variety of forms)
- line graphs
  and perhaps
- scatter diagrams.

The reason for this is that most people will have no difficulty in interpreting these sorts of chart. If you use a more complicated or unusual diagram there is a danger that the reader will not have the time or inclination to interpret the conventions used in the chart and will fail to assimilate the message.

A helpful booklet which matches up the kind of comparison to be illustrated with possible forms of chart is *Choosing and Using Charts* by Gene Zelazny[1] The following table borrows heavily from that booklet.

Table 6.2   Which chart for which comparison?

Sort of comparison	Possible charts
Components as part of a whole	Pie chart. Component bar chart.
Changes in composition of a total	Pie chart. Component bar chart (vertical or horizontal bars).
Sizes of related measurements	Grouped bar charts (vertical or horizontal bars). Possibly isotypes.
Change over time in one or more related measurements	Line graph. Column bar charts.
Change in ranked order of a set of measurements	Paired or grouped bar charts.
Relationship between two sets of measurements	Back-to-back bar charts. Scatter diagrams.

[1] Published by Video Arts, New York, 1972

6.4 How to draw charts. General guidelines

If a chart is to communicate effectively, the reader must find it satisfying to look at and easy to understand. The following guidelines apply to all charts:

1 Make sure your reader is in no doubt about:

- the kind of objects, events, people or measurements represented
- the units used
- the geographical coverage
- the time period covered
- the scale of measurements
- the source of the data
- how to interpret the chart.

All this information should appear round the edges of the chart; that is, in the headings, the labels on the axes and, if necessary, in a clear key (but see **2** below on the subject of direct labelling).

2 Make it easy to read. This means:

- make it no bigger than is necessary for clarity: it is much harder to assimilate information from a large chart than from a small one (A4-size is usually much too big). This can be demonstrated by giving someone an A4-sized chart to look at. The first thing he or she will do is to hold it at arm's length to reduce its effective size.

- keep it the right way up: it is distracting to have to rotate a document through 90 degrees in order to read it; keep all the ancillary markings like axis labels the right way up too

- label pie chart sections, component bar charts and line graphs directly rather than using an explanatory key at the bottom or side of the chart: continual reference to a key interferes with the reader's prime task of assimilating the main features of the chart

- include important numbers (for example, the percentages represented by key slices) on the chart if this can be done without loss of visual clarity

- use clearly differentiated shadings, colours or symbols to discriminate between different

segments of pies or component bar charts and between different lines.

3 Make it easy to understand. This means:

- the conventions used in constructing the graph should be easy to interpret (for example, to represent quantities of different magnitudes use bars of equal widths but varying lengths: the non-specialist reader should not be asked to interpret both changes in length and changes in width)

- each chart must be accompanied by a clear verbal summary of its salient features. The verbal summary should be as close as possible to the chart and direct the reader to the chart (as in 'Figure X, below' or Figure Y, opposite'). It is helpful to quote directly the important numbers on the chart in the verbal summary: this helps to link the chart and the numbers in the reader's memory. But the summary should only comment on a few important points, it should not be a blow-by-blow account of each feature

- do not try to include too much information on a single chart: the reader will find it easier to follow a story built up using a series of simple charts than to disentangle a single complicated diagram.

Figure 6.4, taken from the *Employment Gazette* in June 1993, illustrates many of these points.

Minor improvements might be suggested (for example, more of the percentages printed on the chart might have been quoted in the verbal summary, the titles of the three pies could be clearer and at least one chart could have been labelled directly) but note the following virtues:

1 The information needed to understand it is all provided clearly on the chart:

- we are dealing with economic activity of women in autumn 1992 (that is whether they are in employment, ILO unemployed or economically inactive)

- they are analysed by activity of husband or cohabiting partner aged under 65

- the data were from the Labour Force Survey (the whole page is headed LFS helpline).

2 It is easy to read - that is:

- it is compact and the right way round (no need to twist the book to read the chart)

- the percentages represented by each slice appear on the chart, although they cannot all be read on a black and white photocopy of the chart

- the slices of pie are differentiated by colours which are not likely to be confused visually even when photocopied into black and white.

3 It is easy to understand - that is:

- the conventions used in constructing the chart are easy to interpret: each pie is the same size and each is divided into three slices whose magnitudes represent the percentage of women falling into three defined categories (the fact that each pie represents a different total number of couples can easily be read from the labels below each pie)

- the verbal summary concentrates on a single feature, namely the difference in the unemployment rates for women whose spouses were in employment and ILO unemployed.

Figure 6.4 Economic activity of women (Great Britain, autumn 1992, not seasonally adjusted)

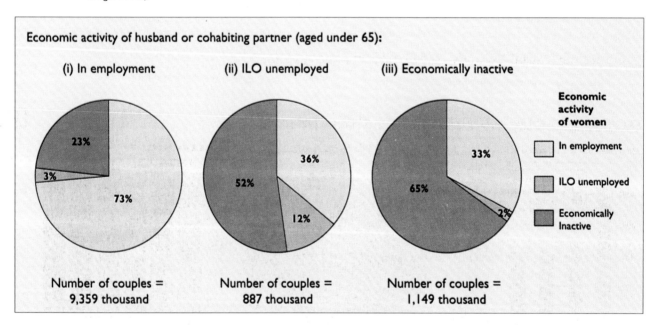

Verbal Summary to Figure 6.4

The three charts shown in Figure 6.4 show the economic activity of women when her husband or cohabiting partner is aged under 65 and (i) in employment, (ii) ILO unemployed or (iii) economically inactive. The charts show that 12 per cent of the wives/cohabiting partners of ILO unemployed men are themselves ILO unemployed compared with only 3 per cent of the wives/cohabiting partners of men in employment.

Charts for publication are often drawn by a professional, a practice strongly recommended whenever possible particularly where a particular house 'look' is essential. Unfortunately not all professional draughtsmen, illustrators and graphic designers know how to draw good statistical charts and the wide availability of graphics packages can tempt the inexperienced presenter into using a variety of unhelpful embellishments. If you decide to design your own charts, you will find useful guidance in the books on charts listed in Appendix B. Cleveland's book *The Elements of Graphing Data* contains a helpful list of the principles of graph construction (though it includes some more specialised graphs than those recommended here).

The three basic principles (charts should be clearly and completely labelled, easy to look at, and easy to understand) apply to all charts. Their application to the four types of chart recommended for non-specialist audiences (pie charts, bar charts, line graphs and scatter diagrams) is discussed in detail in Chapter 7 together with statistical maps and some non-standard chart forms.

6.5 Groups of charts

Groups of charts are often used to good effect by journalists. In publications like the *Economist* and in the business section of quality newspapers, quantitative information is frequently displayed graphically. A number of charts, often of different designs, may be used to illustrate sets of related data: catchy captions are often used and the basic chart may be enlivened in order to amuse (and hence interest) the reader. See Figure 6.5 (from the *Financial Times, March 9 1994*).

When writing for the general public, such tricks are extremely useful and may certainly be used so long as:

- the data are displayed clearly and honestly
- your graphics designer is sufficiently talented to execute the design well.

Figure 6.5 Social Security: the claims on state spending

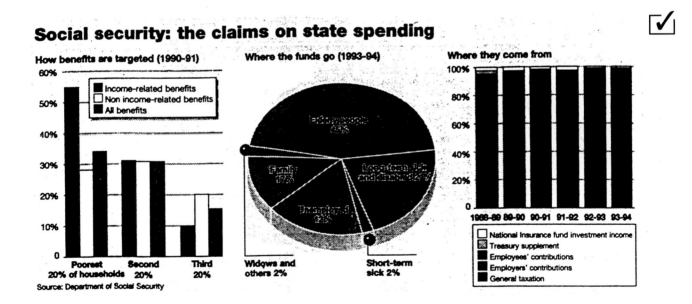

6.6 Three dimensional charts

We do not recommend adding a third dimension to charts for the general reader. True three-dimensional charts, with different variables plotted on three different axes, almost inevitably include too much data for the reader to assimilate easily even if the reader is able to read the scales accurately.

Consider for example Figure 6.6 from *New Scientist*. This is an attractive three-dimensional bar chart but it is not simple. Much more effort is required to read and interpret it than for a two-dimensional chart. The scale is also difficult to judge. Can you tell what the average depth and salinity of the Aral sea were in 1991?

Figure 6.6 Unnatural shrinkage: since the 1960s over 70 per cent of the Aral Sea's water has been lost

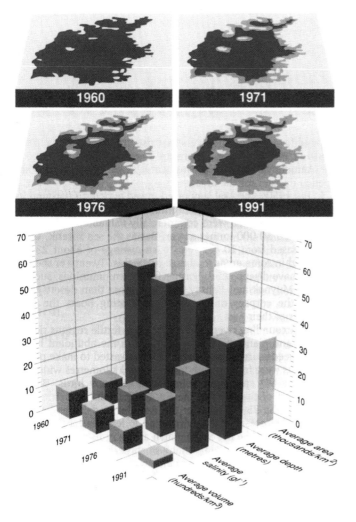

With the advent of desk top graphics, it is becoming relatively easy to give two-dimensional charts a three-dimensional appearance. Pie charts with thickness added are increasingly popular. Adding depth or thickness may seem to enhance the appearance of a chart and make it look more impressive, but it is not recommended. There are three drawbacks:

- the third dimension adds to the complexity of a chart and runs counter to the basic rule of keeping charts simple

- it is often difficult to judge the absolute size or height of lines, bars or segments and to relate them to the scale

- the bars, lines or segments are not drawn to scale; they are distorted to create the three-dimensional effect. You cannot be sure that the readers are seeing what you intend them to see.

The following pie and bar charts illustrate the drawbacks.

The grouped bar chart on retirement pensions in Figure 6.7 gives the clear message that there are more women pensioners than men, but this impression is exaggerated by the way the bars are overlapped. The reader sees more of the red bars representing women and has to work to correct this initial impression. Look at the chart quickly and estimate the ratio of men to women pensioners aged 90+? Then try to work it out more exactly using the scale. Most readers will substantially overestimate the number of women relative to men.

There are in fact about four times as many women as men in this age group, but Figure 6.7 shows seven or eight times more red than black. The true ratio is much clearer in Figure 6.8, where the chart has been re-drawn without the added thickness.

Figure 6.7 Retirement Pension: Age of pensioner at September 1992

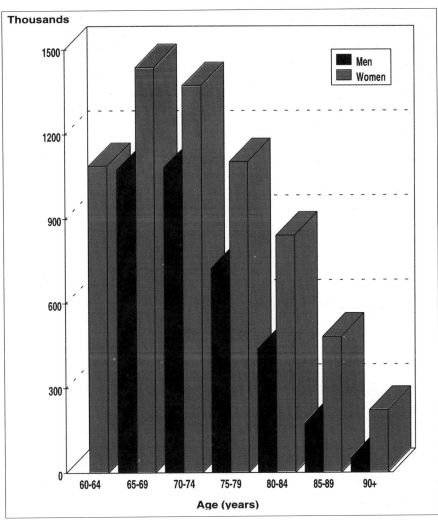

Thousands

| | |
| Men | Women |

Age (years)

Figure 6.8 Retirement pensions (thickness removed)

Thousands

| | |
| Men | Women |

Age (years)

Judging the relative size of components is particularly difficult in pie charts with added thickness. However carefully the pie chart is drawn, the effect of adding thickness is to make nearer segments appear larger than they are. The main problem is that the reader is expected to ignore part of the chart i.e. the height of the nearest segments and compare only the areas within the circle. However casual readers will not do this and may be misled. In Figure 6.9 the yellow component appears, correctly, to be the biggest segment but its size is exaggerated because all the height of this segment is visible and shaded yellow. The blue component at the back appears much smaller because no height is visible.

An additional problem with adding thickness to pie charts is that the circle is sometimes twisted into an ellipse. This distorts the angles and areas of the segments. In Figure 6.9 the pie is clearly elliptical rather than circular and the 'acquaintance' segment has been squashed in the process. The angle of the centre of this segment should be 105 degrees, but as drawn, it is less than 90 degrees. In contrast the 'strangers' segment is expanded from about 115 degrees to 120 degrees. Far less green is visible than blue leaving the impression that rapists are far more likely to be strangers than acquaintances when in fact the difference is quite small. Overall the design of Figure 6.9 conveys the correct message about the ranking of the components, but the differences between them are exaggerated.Figure 6.10 is more acceptable: the chart is nearly circular and very little height has been added. But does the added thickness help the reader in any way?

Figure 6.9 Relationship between rapist and victim 1988

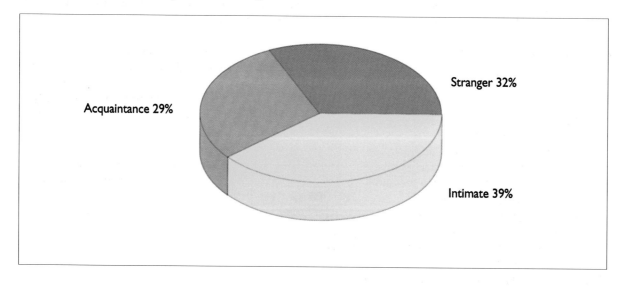

Figure 6.10 One parent benefit

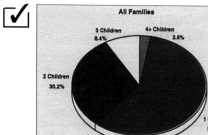

6.7 Special charts

This category is taken to include charts used by specialists (among themselves) and special charts designed to communicate with a non-specialist audience.

The three basic principles described in section 6.4 still apply to these graphs; the only difference is that a greater, more specialised store of background knowledge is assumed to be shared by the producer of the chart and the reader. Specifically, the reader is assumed to be familiar with more complex graphical conventions than the simple use of heights or areas to represent relative magnitudes.

Common examples of graphical conventions familiar to specialists are:

- the use of logarithmic scales by economists

- cumulative sum, or 'cusum' charts, in which the variable of interest is represented by the slope of the graph rather than its height; such charts are useful for detecting systematic changes in processes with a strong random component

- scatter diagrams of residuals, used by statisticians to assess the goodness of fit of a regression model.

Among specialists, writing for specialists, such techniques may be used safely and are often far more effective for particular purposes than conventional bar charts, pie charts or line graphs. Moreover the specialists we refer to here are not necessarily statisticians or economists but could be, for example, accountants, engineers or military officers who are experts in fields served by the statistics. Our reason for not dealing in more detail with specialist charts is not that we do not recognise their value but simply that their lack of universal applicability puts them outside the scope of this book.

A number of graphical methods (many of them recently devised) may be used to explore data. Once the techniques have been mastered and their properties thoroughly understood there is no doubt that they can provide insights into data more easily than other methods. The danger is that in such circumstances, the same graphical technique may be used in presenting conclusions to an audience which has not mastered the properties of the unfamiliar technique. If this happens communication fails altogether.

There are of course circumstances in which the properties of a particular graphical device are so useful that it is worth teaching them to the audience before using this form of graph to present your conclusions. For example, you might wish to demonstrate that two series of data, of very different magnitudes, have grown at approximately the same rate over a 15 year period. This can be demonstrated elegantly and economically using semi-logarithmic graph paper - but only if the audience is aware that the use of a logarithmic scale has the effect of showing constant percentage growth as a straight line whose slope depends on the growth rate (and not on the absolute magnitudes plotted). To explain the properties of semi-logarithmic graph paper adequately to a non-specialist audience will probably require about half a page of text and simple worked example. Whether this amount of effort and distraction from the main report is justified by the effectiveness of the final graph can be judged only by the author. It may be; but it is important neither to underestimate how much effort is required to master the properties of an unfamiliar device, nor to underestimate the discouragement experienced by those who are presented with a chart which they cannot understand.

The process of educating an audience to understand an effective but unusual type of chart is more likely to be worthwhile when you are producing a series of periodic reports to a limited (though not necessarily technical) audience than when you are writing a one-off report for a wide audience.

Two kinds of special chart designed for non-specialist audiences, isotypes and maps, are discussed in Chapter 7.

7 Effective charts for general use

In Chapter 6 the following charts were recommended for communicating with non-specialists:

- pie charts
- bar charts (in a wide variety of forms)
- line graphs

and, perhaps,

- scatter diagrams.

This chapter explains briefly how to draw each type of chart and gives examples of both good and bad practices.

All the charts are reproduced in their original colours and style as closely as possible. This allows the reader to examine the wide range of possibilities open to the statistician with access to colour printing and desk-top publishing. These tools can enhance the appearance of any chart and may make it easier to use. However they also offer new opportunities to create bad charts, where three dimensional presentations distort the message, rainbow colours add to the complexity of the chart or the graphic design obscures the message completely.

We have not found any reported research on the use of colour in charts. There is, however, a clear guiding principle: colour should aid interpretation not add to the complexity of a chart. Two additional points are: colour charts should be designed so that they can be photocopied successfully in black and white; red and green are an undesirable combination given the prevalence of red-green colour blindness.

The comments and criticisms of charts in this chapter are not dependent of the issue of colour. It is always possible to design an effective *simple* chart in black and white and only simple charts are recommended in what follows.

7.1 Pie charts

Pie charts can be used effectively to illustrate:

a the composition of a whole, that is, the relative sizes of different component;

b the difference in composition of two or more related whole.

A pie chart is basically a circle divided into segments in the same way as a cake is divided into slices. The segments vary in size exactly proportional to the components they represent. This is done by calculating the angle at the centre of the cake (as x per cent of 360°. In the pie chart shown in Figure 7.1, 72 per cent of the contribution to global warming came from carbon dioxide so the largest slice has an angle of 259 degrees (72 per cent of 360 degrees) at the centre of the pie; 10 per cent were from methane so that slice of the pie has an angle of 36 degrees at the centre, and so on.

In order to make the composition of the pie as memorable as possible it is helpful to arrange the slices of a pie chart in some natural order. This will often be in decreasing order of size, or importance, with the largest slice positioned against the 9 o'clock line of the chart (or if preferred, the 12 o'clock or 3 o'clock line).

Figure 7.1 Relative direct contribution to global warming from 1990 global emissions of greenhouse gases

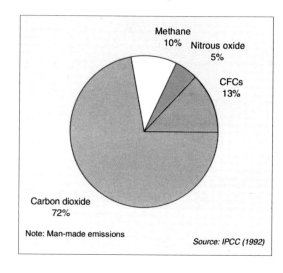

It is counter-productive to include too many slices in a pie chart. You may expect your reader to show an intelligent interest in four or five categories but after that confusion and boredom will set in. The pie charts in Figures 7.2 and 7.3 do not work for this reason. Figure 7.2 has ten segments in seven different colours and because of the number of categories shown, the producer of this chart has had to use a separate key to explain the shadings rather than label the chart directly. It does not photocopy successfully into black and white because most of the colours are equally dense. The white dividing lines show up clearly but the rest is an unhelpful black muddle. Figure 7.3, which is in black and white, uses hatching to differentiate the segments. The variety and type of hatching make this visually unattractive and this presentation can usefully be contrasted with Figure 7.5.

Figure 7.2 Extract from British Council annual report 1991/92

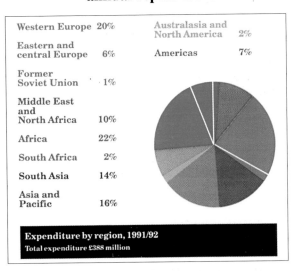

Western Europe 20%	Australasia and North America 2%
Eastern and central Europe 6%	Americas 7%
Former Soviet Union 1%	
Middle East and North Africa 10%	
Africa 22%	
South Africa 2%	
South Asia 14%	
Asia and Pacific 16%	

Expenditure by region, 1991/92
Total expenditure £388 million

Figure 7.3 Staff working on Next Steps lines by department at 1 April 1993

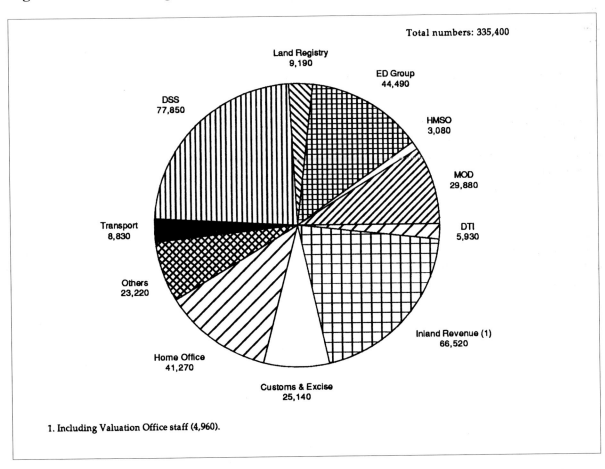

Total numbers: 335,400

Land Registry 9,190
ED Group 44,490
DSS 77,850
HMSO 3,080
MOD 29,880
Transport 8,830
DTI 5,930
Others 23,220
Inland Revenue (1) 66,520
Home Office 41,270
Customs & Excise 25,140

1. Including Valuation Office staff (4,960).

A possible exception to the general rule that the number of slices should be limited to four or five, is if you wish to convey a message like 'category A accounted for over 70 per cent of the total while the other 10 categories between them accounted for less than 30 per cent'. The pie chart in Figure 7.4 clearly conveys the message that in both 1982 and 1992 most of the European Community's expenditure was on agriculture and fisheries.

If a comparatively small slice of the pie is of particular interest, that slice may be shown lifted out of the rest of the pie as in the two examples in Figure 7.5. Here the message being illustrated appears as a caption on the chart: 'A lecturer's duties leave little time for research'. This approach can be used to good effect whenever a chart illustrates a single clear message.

Figure 7.4 European Community expenditure: by sector, 1981 and 1992

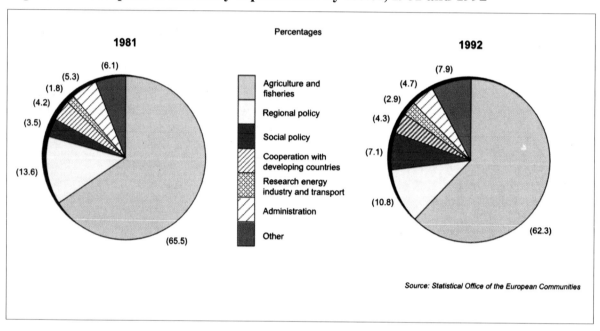

Figure 7.5 Distribution of a college lecturer's time

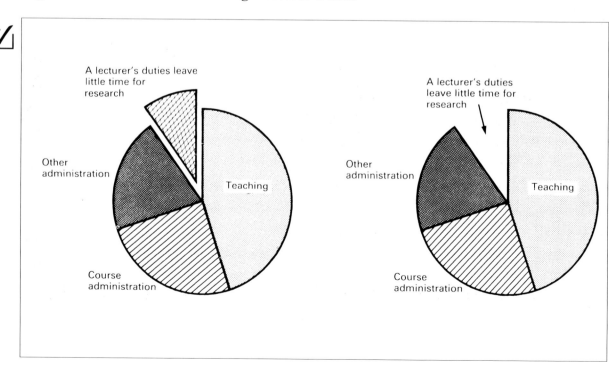

The reader's attention can also be focused on a comparatively small but significant component of a pie chart by using shading or colour as in Figure 7.6. But enlarging the segments as in Figure 7.7 may be misleading.

Figure 7.6 Radiation exposure of the population: by source, 1991

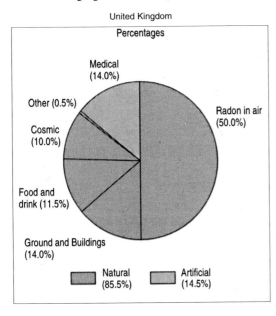

Before moving on to discuss bar charts, it is worth mentioning another possible embellishment of pie charts. First the pie may be drawn as a symbol appropriate to the total whose composition is illustrated (so long as it is circular) for example a coin, a cake or, indeed a pie as in Figure 7.8. However here the design is so strong that the message of the chart is obscured and this is not an example of good practice.

Figure 7.9 also has an interesting graphical design which makes an attractive picture but does not work as a chart. Here the relative size of the segments is unclear, because the car is not circular. In any case there are far too many segments for any message to emerge.

Two or more pie charts can be used to compare differences in the composition of different totals or the same total at different points in time. A number of things must be considered to do this successfully:

- arrangement of segments
- size of pies
- number of pies.

Figure 7.7 Extract from Race and the Criminal Justice System

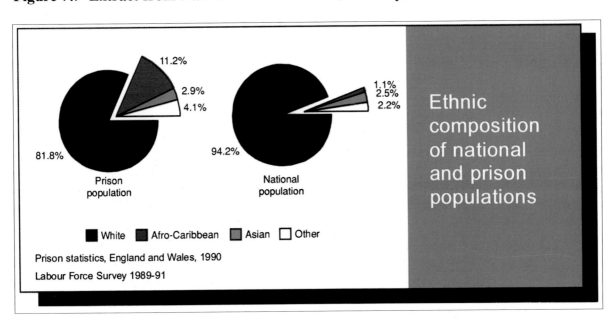

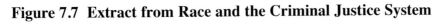

Figure 7.9   Cars on road at end of 1992 by year of registration

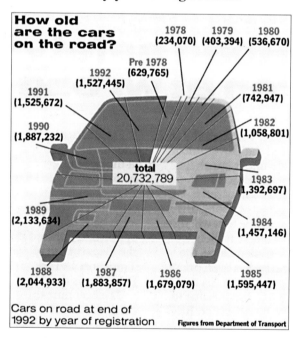

The segments should be arranged to make the comparison easy. Both charts should start from the same baseline and segments must appear in the same order in all the pie charts. Figure 7.10 does this by ranking the segments in both charts in 1990 order of size starting at 3 o'clock. By contrast Figure 7.11 starts both charts from a 3 o'clock baseline but orders the segments differently and in no apparent order.

Figure 7.10 illustrates a second point which is important when the composition of two or more pies is to be compared: the pies are the same size, even though they represent different total amounts (1,343 million tonnes in the first case and 2,160 million tonnes in the second). Some statistical text books recommend that in cases like this the area of each pie should be proportional to the total represented. However, there is ample evidence[1] that most people are unable to estimate the relative areas of different circles accurately, and so carrying out the complicated calculations needed to scale the areas of two or three pies to represent varying totals correctly merely wastes time. Compare Figures 7.11 and 7.12 which appeared as a pair in the *Digest of UK Energy Statistics* 1993.

[1] For example: MEIHOEFER HJ. The visual perception of the circle in thematic maps; experimental results. *Canadian Cartographer* 1973, vol 10, pp63-84

Figure 7.10 Freight transport by mode: 1955 and 1990, GB

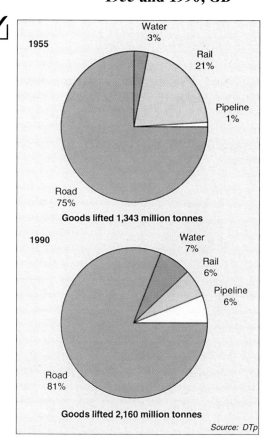

Goods lifted 1,343 million tonnes

Goods lifted 2,160 million tonnes

Source: DTp

Figure 7.11   Energy consumption by type of fuel

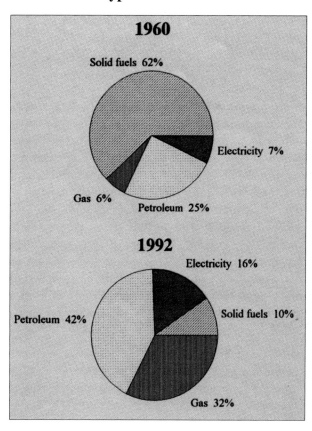

Figure 7.12   Energy consumption by final user

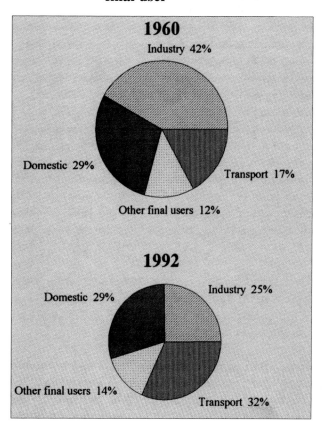

The areas of the two pies in Figure 7.12 are different and it would be reasonable to assume that they were drawn to reflect differences in the totals represented. But how much bigger is the top pie? Very few people can make such judgements accurately (and compare the percentage of energy consumed by different final users as well). Even though the ordering of the slices is unhelpful, most people will find Figure 7.11 easier to deal with. Here the relative sizes of different slices can be compared immediately (the reader being directed to the slices of greatest interest by the verbal summary, of course) and the increase in the share of petroleum (from 25 to 42 per cent) can be seen clearly. In passing, the reader may like to note that the scaling in Figure 7.12 was unintentional in the original publication and was also incorrect; energy consumption grew between 1960 and 1992.

Figure 7.13 shows three pies on the same chart. This is probably the largest number of pies which should be presented on a single chart. Occasionally four pies can be effectively displayed in the same diagram but never more. What patterns do you see in Figure 7.14? To compare the composition of more than three totals, component bar charts should generally be used. (See section 7.2)

Figure 7.13 Type of transaction - principals' turnover

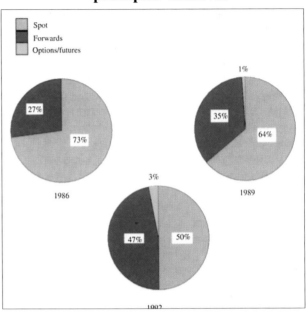

Figure 7.14 Highest qualification of the ILO unemployed, by age: spring 1991 and spring 1992

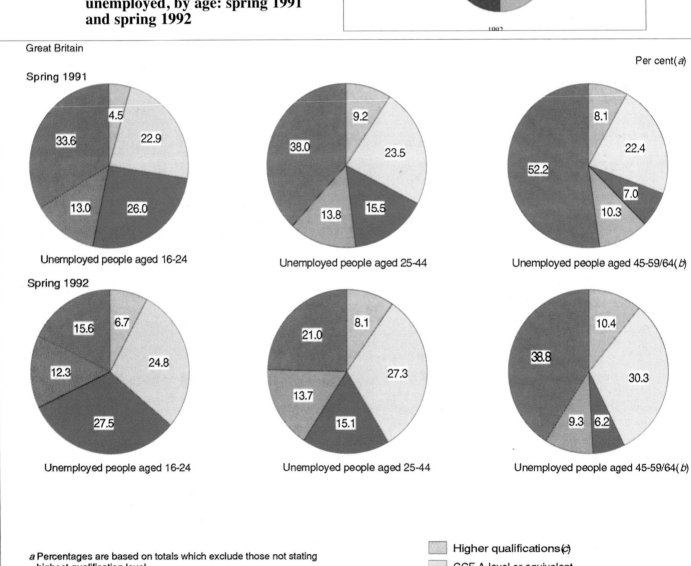

a Percentages are based on totals which exclude those not stating highest qualification level.
b The upper age limit is 64 for men and 59 for women
c Above GCE A-level or equivalent: see footnote to table 14
d Includes YTS certificate

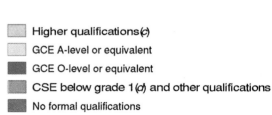

Higher qualifications(c)
GCE A-level or equivalent
GCE O-level or equivalent
CSE below grade 1(d) and other qualifications
No formal qualifications

Source: LFS estimates

7.2 Bar charts

Bar charts are very versatile: they can be used to illustrate the magnitudes of related measurements, the composition of related totals, changes over time in the composition of a single total and also the degree of correlation between two measurements. A similar chart, the histogram, can be used to illustrate the frequencies with which different measurements occurred in a set of data.

Before discussing different uses of bar charts in more detail, let us dispose of one minor issue.

Horizontal or vertical bars?

Most bar charts can be drawn using either horizontal or vertical bars. In terms of the ease with which a chart is understood, the orientation of the bars appears to make little difference. It is however, helpful to label bars and sections of bars directly rather than by means of a separate key and this may be easier to do on horizontal bars than on vertical ones. The following pages contain several examples of both horizontal and vertical bar charts, indicating that either sort can be used effectively (or ineffectively): the orientation of the bars is less important than the clarity with which the final chart matches its message.*Sizes of related measurements: bar charts and grouped bar charts*

Simple bar charts can be used effectively to show the number of items in specified categories, the percentage of items in specified categories, or incidence rates in specified categories. In all cases bars of equal widths are drawn with lengths proportional to the measurement being illustrated.

In Figure 7.15 the bars are drawn horizontally, primarily for ease of labelling each bar. When using horizontal bars it is advisable to put the labels alongside and to the left of each bar if possible. If names of different lengths are written to the right of or inside the bars, the eye no longer registers a clear profile of the relative lengths of each bar. See for example 'Professional (62)' and 'Personal protective services (74)' in Figure 7.16.

Figure 7.15   Average household size: by ethnic group of head of household, 1991

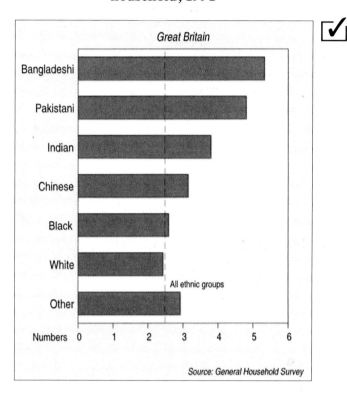

Figure 7.16   Percentage of employees or self-employed working in their own home: by occupation

Great Britain, spring 1993, not seasonally adjusted

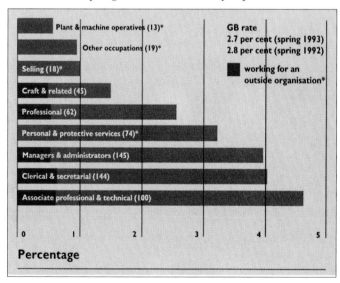

The difficulties of labelling a vertical chart are clearly illustrated in Figure 7.17 from *Autocar and Motor*. The words do not fit when written horizontally and have been written vertically instead. It is distracting to have to turn the book through 90 degrees to read the labels.

Figure 7.17   Features you'ld change

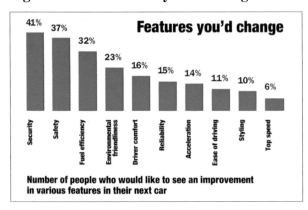

Unless there is a very strong natural order for the groups such as age or time, the bars should be arranged in order of size. The largest category can be at the top as in Figure 7.15 or the bottom as in Figure 7.16; either arrangement can be used effectively. Where the 'natural' order is a standard classification such as region or industry, there is less justification for using the standard order in a chart. The classification may not be familiar to the reader and will not help them use the chart. Can you tell from Figure 7.18 which region has the highest proportion of males consuming alcohol above sensible limits? Look instead at Figure 7.19 where the chart has been re-ordered. The pattern for males is instantly apparent even if the pattern for females is not. However we can also see clearly that there is no correlation between the rates for males and females by region.

Figure 7.18   Consumption of alcohol above sensible limits[1]: by sex and region, 1990

1 Persons aged 16 and over consuming 22 units or more for males, and 15 or more units for females, per week.

Source: General Household Survey

Figure 7.19   Consumption of alcohol above sensible limits[1]: by sex and region, 1990 (re-ordered)

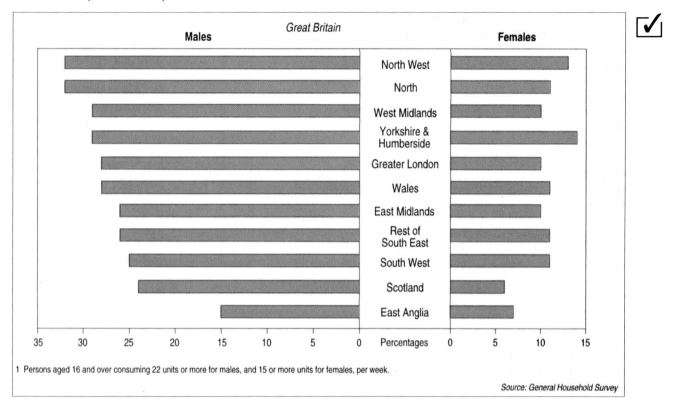

1 Persons aged 16 and over consuming 22 units or more for males, and 15 or more units for females, per week.

Source: General Household Survey

Change over time: column bar charts

Change over time in a specified measurement can be illustrated using either a series of bars (usually drawn vertically) or as a line graph. A column bar chart is preferable when comparatively few periods are plotted or where the measurement refers to an activity completed within distinct time periods. Figure 7.20 from *The Economist* illustrates the decline in profitability of Sotheby's and Christie's in the early 1990s under the title of 'Going, going, gone'. A line graph (discussed in section 7.3) is likely to be more effective where a large number of readings are displayed or where there is a carry-over effect from one time period to the next and you wish to give an impression of continuous movement. (For example, the number of births per year might be shown as a series of bars while the total population recorded year by year might be plotted as a continuous line.)

Figure 7.20   Going, Going, Gone

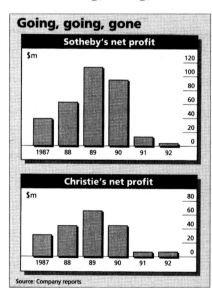

The Economist June 26th 1993

Graphical devices such as lines, colour and shading can be used to help the reader make comparisons and thus get more information from the chart. For example lines can be added to show averages, scales and trends. These devices can be particularly useful if the bars cannot be ranked in size order. However including too many lines, or drawing them too heavily will give the chart a cluttered appearance and this should be avoided. The use of an 'average' line is illustrated in Figure 7.15. Scale lines are included in Figure 7.16 but are drawn a little too heavily. Figure 7.21 includes a trend line, which is now common in *CSO first releases*[1].

Colour and shading can also be used to highlight bars which the author thinks will be particularly interesting. Figure 7.22 uses a different colour to highlight 'England and Wales' but shading would work equally well in a black and white chart.

---

[1] *CSO first releases* are press releases from the *Central Statistical Office* containing the latest statistics.

Figure 7.21   The balance of payments current account

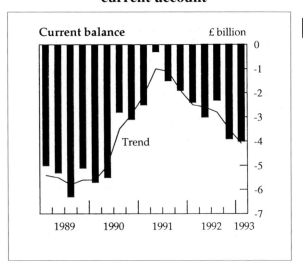

Figure 7.22   The risk of being burgled: international comparison 1988/91

■ The risk of being the victim of burglary in E & W was one of the highest in Europe

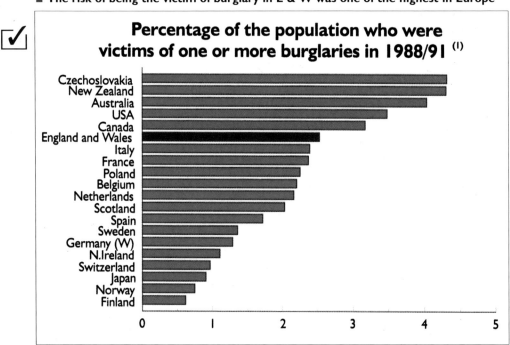

Grouped bar charts

Grouped bar charts can be used to compare the distribution of a set of measurements at different points in time, in different geographical areas or according to some other factor such as sex or age. In these charts each bar is accompanied by one or more others showing the magnitude of the same measurement at a different time, in a different area or for a different sex or age group. It is important to group such bars in the most effective way. Compare, for example, Figure 7.23 from *The UK Environment* with Figure 7.24 which we have redrawn to group related measurements together. The relative numbers of sheep, cattle and pigs are still clear, but the trends over time for cattle and, particularly, pigs can be examined much more easily in Figure 7.24.

It is usually best to restrict grouped bar charts to two or three bars per group. With four or more, visual impact is lost unless one set of bars shows a markedly different pattern from the others. Figure 7.25, for example, conveys no clear message other than that most goods go by road.

Groups of corresponding bars can be drawn touching each other as in Figures 7.23 to 7.25 or narrowly separated as in Figure 7.26 from the *BP Statistical review of world energy*.

Figure 7.23   Livestock numbers 1980, 1985 and 1990, UK

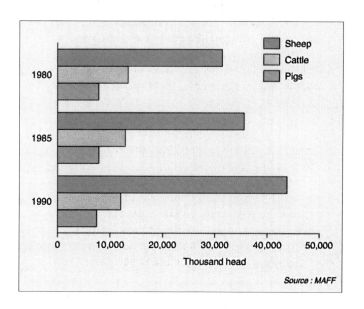

Figure 7.24   Livestock numbers 1980, 1985 and 1990, UK (redrawn)

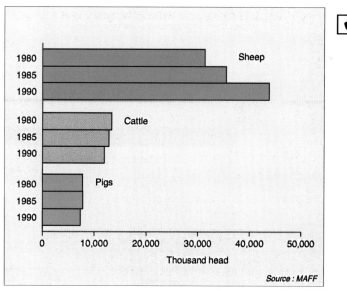

91

Figure 7.25   Goods transport in the European Community 1989 (per cent)

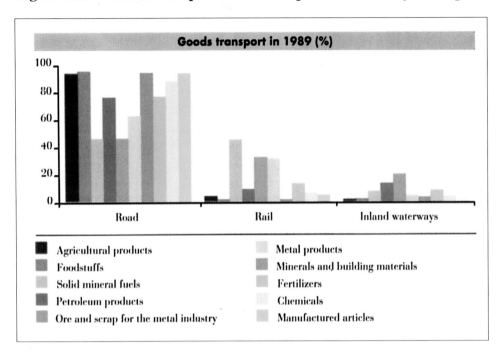

Figure 7.26   World energy consumption per capita

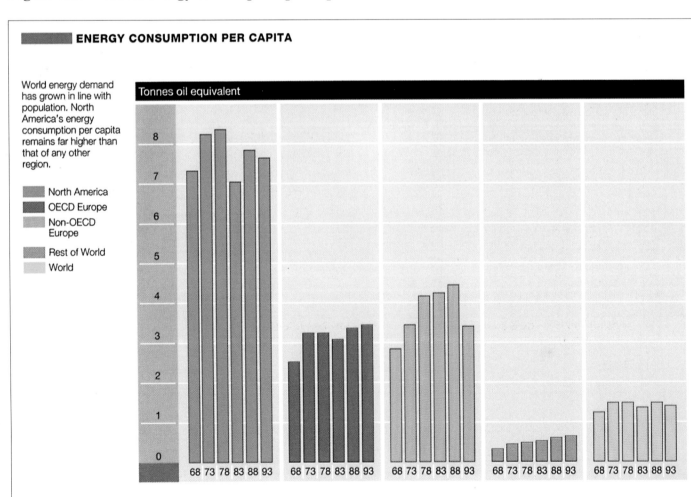

Two, or possibly three, related measurements can also be displayed using overlapping bars. This technique works best if the front bar is consistently smaller than the other(s) as in Figure 7.27.

Sets of bar charts may be an effective alternative to grouped bars. They can be placed side by side as in Figure 7.28 or above each other.

As with pie charts 3 or 4 bar charts is the maximum that can be effectively displayed together. The attractive looking Figure 7.29 is a clear case of overkill.

Correlation between two sets of measurements

Statisticians frequently explore the relationship between two sets of measurements by plotting one measurement against the other in a scatter diagram (see section 7.4). However, non-specialists may not be used to interpreting scatter diagrams and may find a systematic association between measurements easier to see if back-to-back charts are used.

Figure 7.27   Rainfall: potential water resources under normal and drought conditions

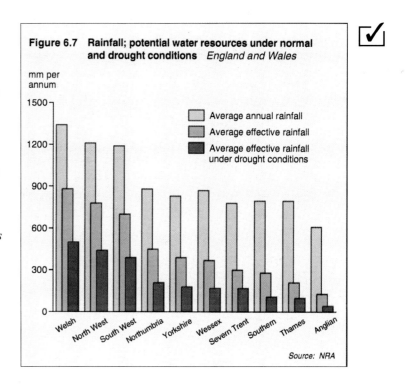

Figure 7.28   Attrition within the Criminal Justice System

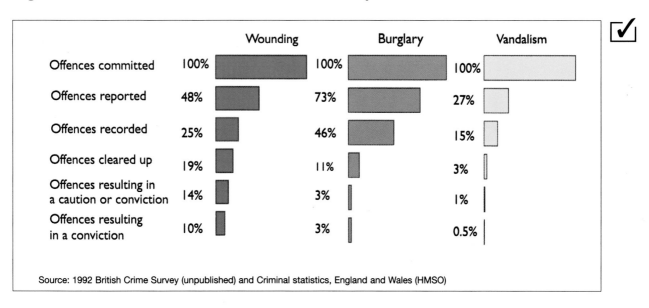

Figure 7.29   Inland water quality and pollution by region

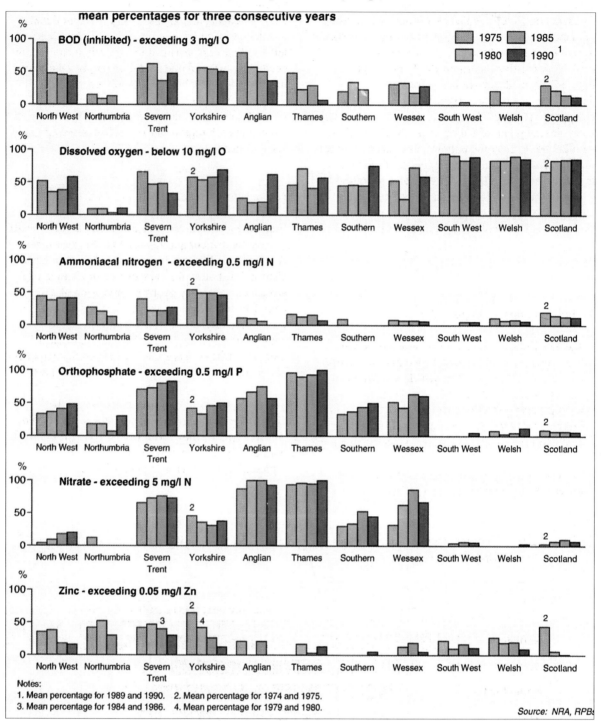

Notes:
1. Mean percentage for 1989 and 1990.   2. Mean percentage for 1974 and 1975.
3. Mean percentage for 1984 and 1986.   4. Mean percentage for 1979 and 1980.

Source: NRA, RPBs

To do this the data are first arranged in ascending or descending order of one measurement. This measurement is represented as a series of increasing or decreasing bars. The second measurement is then displayed as bars, each back-to-back with the corresponding first measurement. For example, in Figure 7.30 the data have been arranged in descending order of the original male claimant unemployment rate (Northern at the top, with a male unemployment rate of 8.4 per cent; the South East, with 2.9 per cent, at the bottom). The inactivity rate for men aged 60-64 in each region is represented by the right-hand bars. The inactivity rates for 60-64 year olds tend to follow the same pattern as their unemployment rates, but with some exceptions. These are highlighted in the verbal summary.

Verbal summary to Figure 7.30

The chart illustrates that, in general, areas with high unemployment also recorded high inactivity rates for 60 to 64 year old men. The Northern region recorded both the highest claimant unemployment rate (8.4 per cent) and the highest inactivity rate for 60 to 64 year old men, 21 per cent. However Scotland, with a higher claimant unemployment rate than the North West, Wales, or Yorkshire and Humberside, recorded a slightly lower inactivity rate than any of these regions.

A complete lack of association between two measurements can also be displayed using back-to-back bar charts, although this is only worth doing if the lack of association is surprising. In this case the left-hand bars are arranged in regularly increasing or decreasing order of one measurement, and the right-hand bars will exhibit a completely irregular profile. Figure 7.19, for example shows that there is no correlation between the proportions of men and women in each region whose alcohol consumption is above sensible limits.

Composition of a total: component bar charts

Either pie charts or component bar charts can be used to illustrate the composition of a total measurement. A pie chart is likely to be more effective than a bar chart in illustrating the composition of a single total since a circle gives a stronger impression of a complete entity than does a bar, but either can be used effectively to illustrate *changes* in the composition of a total. When the compositions of more than three totals are to be compared, component bar charts are likely to work better than pie charts.

There are two ways of drawing component bar charts. The first consists of drawing a series of bars whose height (if drawn vertically) or length (if drawn horizontally) is proportional to the total being represented. Each bar is then sub-divided

Figure 7.30   Percentage claimant unemployment and inactivity rates for males aged 60-64: by region

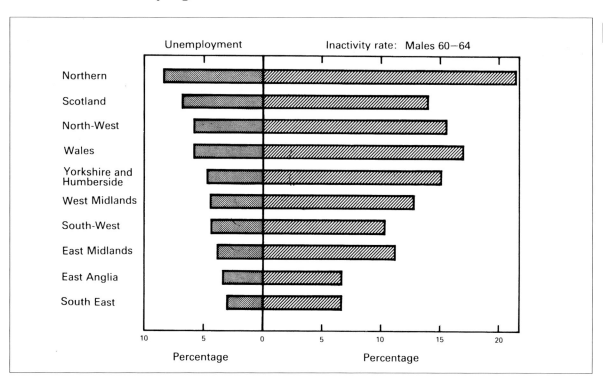

95

into components whose sizes are proportional to the components. The second way is to use a *percentage component bar chart* where all bars are the same length, regardless of the total represented, and the chart is used to emphasise changes in the composition of the total.

The first approach should be used with great care. It is best used only to illustrate a marked change in the relative importance of a single component. For example, if each bar consists of two components, one of which has remained relatively constant while the second component has changed, the relatively constant component will appear at the base of the bar and the verbal commentary is likely to highlight this feature - something like:

'although the total number of people employed has decreased steadily over the last five years, from 450 in 1987 to under 300 in 1992, the number of males has remained fairly constant at about 180. The decrease in the total is due almost entirely to a fall in the number of females employed.'

Figures 7.31 and 7.32 both effectively illustrate marked changes in the relative importance of a single component at the same time as in the composite measurement.

In Figure 7.31 we see immediately that the ethnic minority population is a very young population with about seven times as many people in the 0-15 age group as in the 60 and over group. We also see a marked contrast between the age distributions of people born in the UK and those born overseas: most of the 0-15 group and about half in the 16-29 group were born in the UK, but nearly all of the older age groups were born overseas.

In Figure 7.32 there are two clear messages: the total amount of lowland grassland fell between 1932 and 1984 and the composition of the total changed dramatically. The areas of rough grazing and semi-natural pasture and meadow both declined sharply, while the area of leys and improved grassland increased markedly.

Figure 7.31   Ethnic minority population: by age and whether UK-born or overseas-born, 1987-1989

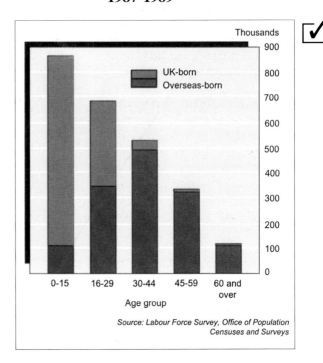

Source: Labour Force Survey, Office of Population Censuses and Surveys

Figure 7.32   Lowland grassland, 1932 to 1984

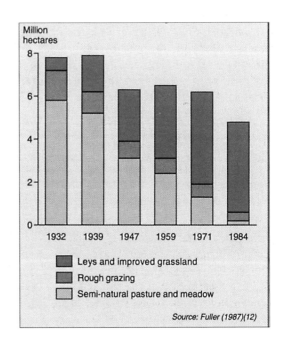

Source: Fuller (1987)(12)

Unfortunately, component bar charts frequently contain too much information to be effective by including both changes in a total measurement and changes in each of the constituent parts. In general the end result illustrates nothing very clearly. Glance at Figure 7.33 and see what you make of changes in the crop area of potatoes and horticulture.

In cases like this the recommended approach is to identify the main message in the data and then choose a chart to illustrate it. If changes or differences in the constituent parts are of greatest importance, a grouped bar chart can be used as, for example in Figure 7.25 (though this is not a particularly good example).

If both changes in total and changes in composition must be highlighted, two simple charts should be considered: for example a simple column bar chart or a line graph (discussed later) to illustrate the change in the total and two (or possibly three) pie charts demonstrating the composition of the total at the beginning and at the end of the period (and, possibly, at a mid-way point).

There are circumstances where a change in total *and* a change in composition can be successfully shown on the same chart but they are rare.

In general, if the composition of a set of totals is to be illustrated, it is safer to use a percentage component bar chart like Figure 7.34 , where all bars are the same length. Where different totals are represented this should be indicated by labelling each bar.

Skill is required in deciding on the order of presentation of the bars and in choosing the order of components within each bar. Where there is no natural order, it is helpful to present the bars in ascending or descending order of one major component. A regular pattern is easier to follow than an irregular one.

Figure 7.33   Crop area by type of crop, 1980 to 1990 UK

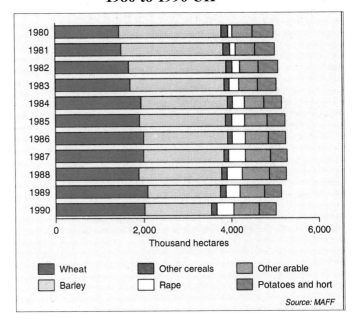

Figure 7.34   Britain's changing shopping basket

In Figure 7.35 a chart from *Europe in figures* is shown in its original form. The same information is shown in Figure 7.36 with the bars rearranged as far as possible in descending order of the 'EUR 12' component. This has been chosen because European agriculture is the focus of the table.

The components of the bars should also be arranged in some natural order if there is one. Where the components have no natural order they should usually be drawn in descending order of magnitude with the largest or most important component at the

bottom if vertical bars are used and at the left-hand side if horizontal bars are used.

Taken together the last two guidelines indicate that where there is no natural order either for the components of each bar or for the bars themselves, the components should be arranged first, with the largest (or most important) component at the base of each bar. The bars should then be arranged to show this component increasing or decreasing across the bars.

Figure 7.35   European agriculture in the world: a major force

Share of world production (%)

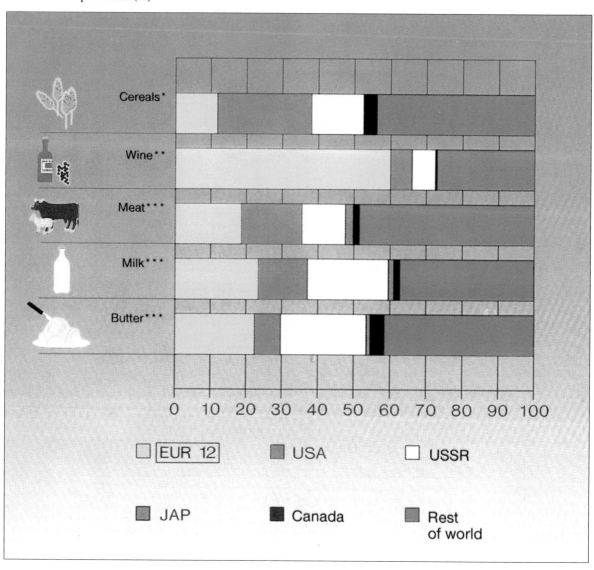

(The exception to this ordering is the 'other' category which remains at the end. This sort of catch-all category often presents problems: it is likely to fall midway through the list if any systematic arrangement of bars is applied, but because it usually consists of a rag-bag of small categories with no dominant component its natural place seems to be at the end of the data. We have not found a tidy solution to this problem, but imposing a systematic ordering on the remaining bars increases the chance of the sequence being remembered.)

However it is wise to explore several possibilities. A different ordering of the bars or components may highlight different comparisons within the data. For example, it is easiest to compare components with a common starting point, particularly those at either end of the bars. A particularly interesting component might thus be placed at one end. Alternatively components which do not vary very much can be placed at the ends of the bars so that the components in the middle have similar starting points.

Figure 7.36 European agriculture in the world: a major force (bars re-ordered)

Share of world production (%)

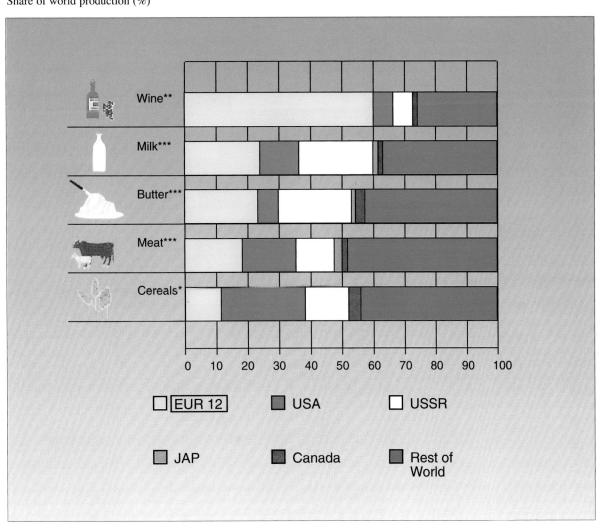

In Figure 7.37 the chart has been simplified further and components of the bars re-ordered. Canada and Japan have been subsumed into the 'rest of the world', because they are so small that the reader is unable to gain any useful information about them from the chart. Other groupings for example 'USA plus Canada', would have been equally useful.

The 'rest of the world' has also been moved into the middle of each bar. The 'other' category is probably the least interesting and here it has been placed in the middle of the bar, where it is most difficult to draw conclusions about the data. The order of the USA and USSR is not clear cut and could easily have been reversed.

Figure 7.37   European agriculture in the world: a major force (bars and components of bars re-ordered)

Share of world production (%)

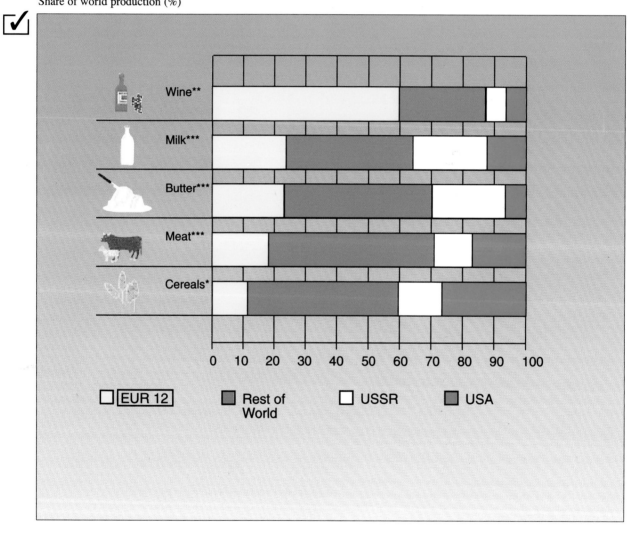

Figure 7.38   Composition of income and expenditure of Buckinghamshire County Council, 1982

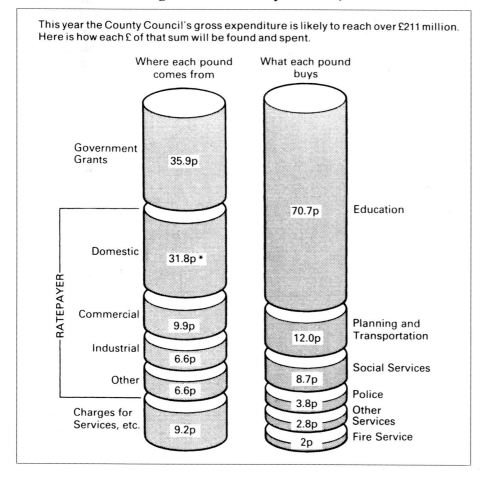

This year the County Council's gross expenditure is likely to reach over £211 million. Here is how each £ of that sum will be found and spent.

Where each pound comes from

Government Grants — 35.9p

RATEPAYER
Domestic — 31.8p *
Commercial — 9.9p
Industrial — 6.6p
Other — 6.6p

Charges for Services, etc. — 9.2p

What each pound buys

Education — 70.7p

Planning and Transportation — 12.0p

Social Services — 8.7p

Police — 3.8p

Other Services — 2.8p

Fire Service — 2p

One common and effective use of percentage component bar charts is in displaying where money comes from and how it is spent. Figure 7.38 shows how Buckinghamshire County Council presented this information to the public in 1982. Note the device of slightly separating the components in order to make it easier to identify individual components. But notice also that the 'Government Grants' at the top of the left-hand stack appears much bigger than 'domestic' because of the extra surface area showing, when in fact the difference in size is only 10 per cent.

Histograms

A histogram is a statistical diagram in the form of a particular kind of column bar chart. Figure 7.39 shows the distribution of radiometric age determinations of samples of rock around the Arabian Shield.

Technically, a histogram is defined as a series of rectangles each of whose bases represents a range of measurements, called a class interval, and whose area is proportional to the number of measurements falling into the corresponding range. For example, in the histogram in Figure 7.39 the fifth class interval covers the age range 4 to 5×10^8 years and there were 52 samples whose radiometric age was determined as lying in that range.

In drawing a histogram, care must be taken over two points. First, the class intervals must be defined without overlap; thus class intervals defined as:

1-2; 2-3; 3-4 etc.

are *not* acceptable since there is no indication whether the measurement 2 is assigned to the first or the second class. Class intervals must be more tightly defined as, for example:

1 and less than 2; 2 and less than 3 etc.

or alternatively:

up to and including 1; over 1 and including 2 etc.

This point need really concern only the collector of the data. In Figure 7.39 it is not clear what convention has been adopted in defining the class intervals, but the shape of the data is perfectly clear to the reader: the majority of samples had ages determined as lying between 5 and 7 x 10⁸ years, and, on either side of this central peak, the age determinations were more or less symmetrically distributed, with the noticeable exception of a cluster of readings under 1 x 10⁸ years.

Figure 7.39 Distribution of radiometric age determination of rocks in Saudi Arabia

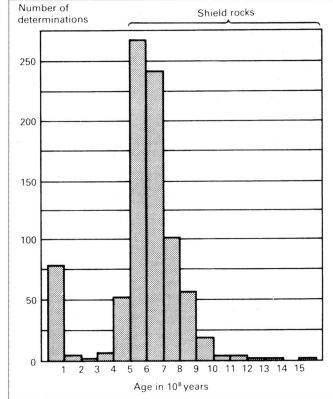

The second point is one of more importance in presenting the data; it concerns the definition of a histogram as a series of rectangles whose *areas* are proportional to the number of observations in each class interval. There is no problem where all class intervals are of equal width; in such cases, since all the rectangles have equal bases, their heights must be proportional to the number of observations in each class. However, in cases where there are class intervals of different widths, the height of each rectangle must be calculated so that the area represents the number of observations. This can be confusing to readers not familiar with the idea.

Presenters avoid this complication wherever possible because bars of different widths can distract the reader from the overall pattern. A possible solution to this problem is to show only the outlines of the histogram. The reader is no longer distracted by the need to interpret bars of different widths, and can concentrate on the profile.

Misuse of bar charts

This lengthy section has illustrated some of the many ways in which bar charts can be used effectively: we have seen bars drawn singly, in pairs, in groups, overlapping, back-to-back, drawn vertically and horizontally, sub-divided into components or drawn as piles of money; the lengths of bars have been proportional to absolute magnitudes, to percentages and to the numbers of items falling in specified categories. Unfortunately this versatility can lead to the misuse of bar charts. Because so much information *can* be shown in a bar chart, there is a temptation to cram too much on to a single chart. Figures 7.40 and 7.41 are two examples of bar charts hopelessly overpacked with information.

When using bar charts, it is extremely important to resist the temptation to 'put it all in the chart in case someone wants to look at it that way'. Always remember your duty to the reader: a chart is provided to illustrate an important point, not as a data store.

Figure 7.40   Pattern of expenditure for different household types, 1989

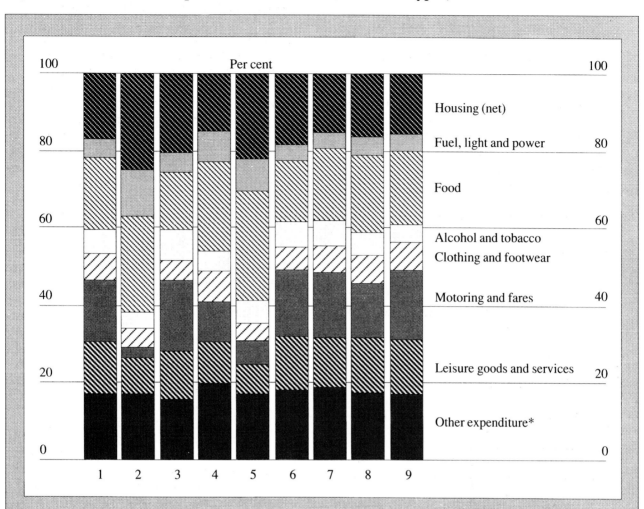

Household composition	No. of cases	Average weekly household expenditure (All items) (£)
1 All households	7410	224.32
2 1 adult retired, mainly dependent on state pensions	618	59.78
3 1 adult, non-retired	859	140.79
4 1 adult with 1 or more children	334	132.40
5 1 man, 1 woman retired, mainly dependent on state pensions	301	101.64
6 1 man, 1 woman non-retired	1442	269.46
7 1 man, 1 woman, 1 child	639	256.42
8 1 man, 1 woman, 2 children	822	284.04
9 1 man, 1 woman, 3 children	280	318.22

*Other expenditure includes expenditure on household and personal goods and services and miscellaneous items.

Note: Percentages are expenditure on commodity or service group as a percentage of total household expenditure

Figure 7.41   Average prison population by length of sentence

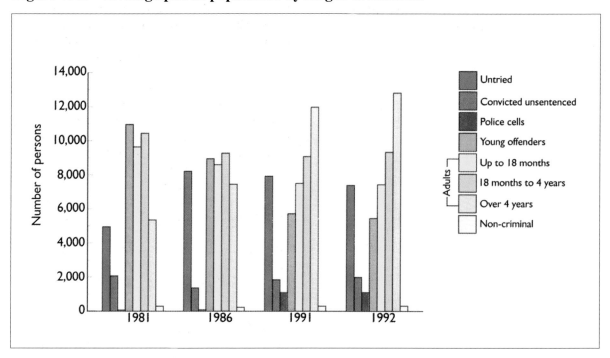

- The sentenced young offender population fell from an average of 10,600 in 1981 to 5,500 in 1992 (5,900 in 1991). The adult population sentenced to over 18 months rose from 13,500 in 1981 to 21,700 in 1992 (21,500 in 1991) with falls for shorter sentence length prisoners.

- The average prison population of female offenders rose from 1,410 in 1981 to 1,580 in 1992 (1,560 in 1991).

Source: Prison statistics, England and Wales (HMSO). The Prison Population in 1992 (Home Office Statistical Bulletin 7/93).

7.3 Line graphs

Line graphs can be used effectively to illustrate the movement of a measurement over time. They are quick to draw and easy to understand. Figure 7.42 clearly illustrates the large reduction in infant mortality over the past 70 years. It also shows that infant mortality fell sharply in the first half of the period, but that improvements since have been less pronounced as low levels of infant mortality have been reached. Similarly 7.43 is very simple, but also has a memorable message; the percentage of live births outside marriage was stable between 1900 and 1960 apart from two small peaks in the war years. But since 1960 the percentage of births outside marriage has increased dramatically.

Figure 7.42   Infant mortality

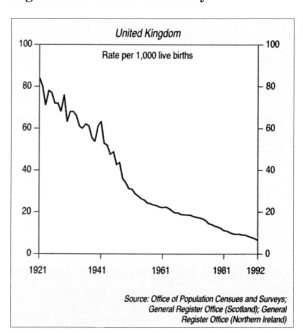

Source: Office of Population Censuses and Surveys; General Register Office (Scotland); General Register Office (Northern Ireland)

Figure 7.43   Live births outside marriage as a percentage of all births

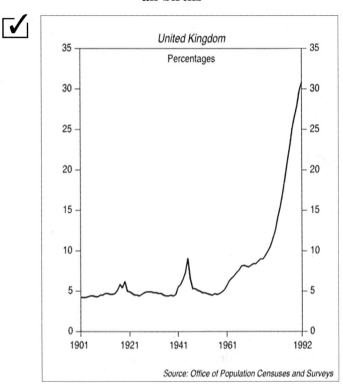

United Kingdom
Percentages

Source: Office of Population Censuses and Surveys

Some data are measurements taken at one point in time, whereas others are averages for a certain period. The author must decide whether these features of the data matter to the reader and present the data accordingly. Figure 44 shows a way of doing this in a chart. The bars represent averages from one data source and are as wide as the survey period (March to May until 1991). They are not joined together because there is no data for intervening periods. The dots represent samples taken at points three months apart and are plotted at the point in time when the sample was taken. They are joined by a line to show the trend over time. Strictly speaking, such a line is an artefact and could be misleading if the data are seasonal or volatile.

The presentation of the data in Figure 7.44 is technically correct, but focusing too strongly on technical aspects may make a chart over complicated for a lay audience.

In our remaining examples of line graphs, the points represent the total or average figures over given intervals (usually years). Once again the lines joining the points are, strictly speaking, artefacts, and purists may prefer to use bar charts instead. They take longer to draw, however, and often result in a fussier, less effective presentation (imagine, for example, a bar chart version of Figure 7.43).

Figure 7.44   Employees: LFS and employer survey-based cstimates 1984-93

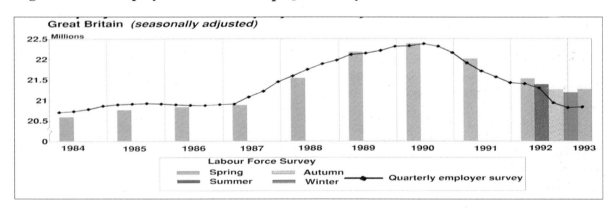

Great Britain *(seasonally adjusted)*
Millions

Labour Force Survey
Spring   Autumn
Summer   Winter   ●—— Quarterly employer survey

A point which needs care arises when you wish to break and expand the vertical scale on a line graph. This usually occurs when all the observations lie above a comparatively high value. For example, the author of Figure 7.44 wished to show changes in employment between 1984 and 1993. All the observed values are between 20 million and 22 million so that, if the vertical axis had been calibrated from 0 to 22 million there would have been an expanse of blank space at the base of the chart - as there is in Figure 7.45.

The featureless expanse of colour or blank space at the base of the chart may seem unattractive. Actually, it serves the important purpose of enabling the reader to assess at a glance the *proportional* change in employment. If the reader understands (or has been clearly told) that the proportional change is small and the exact pattern of change is of particular interest, the vertical axis may be broken and expanded as in Figure 7.44. The broken scale must be indicated with either a zig-zag or break in the scale line and ideally a narrow gap in any background shading. Breaking the scale is a device for occasional use in special circumstances and must be employed with care.

Comparison of growth rates

Line graphs can also be used effectively to compare growth or decline in two or more series of related measurements over the same period, but here great care must be exercised. It is important that the chart should convey a clear visual message and the temptation to put too much on a single chart must be resisted.

Compare for example the clear message of the bottom left-hand chart in Figure 7.47 about whole and skimmed milk with the tangle of lines in the top right-hand chart about meat and fish consumption.

Figures 7.46 and 7.48 are much worse examples; they convey little impression beyond a general tangle. In both cases the information would be stored more effectively in a reference table, although a chart could be used to support commentary on a few of the series displayed.

Figure 7.46   NO₂ concentrations at selected sites, 1976 to 1980

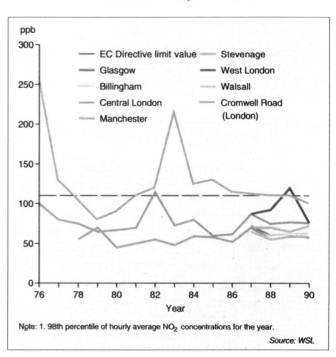

Figure 7.45   Workforce and workforce in employment

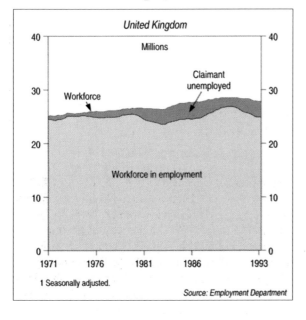

Figure 7.47 Changing patterns in the consumption of food at home

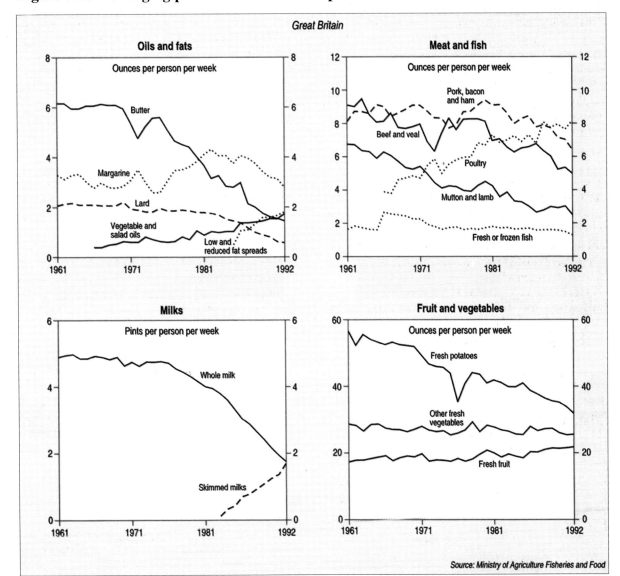

The sort of messages which are likely to lend themselves to effective line graphs are:

- a number of series moving in parallel
- a small number of series moving in parallel and one (occasionally two) series showing a markedly different pattern from the rest.

Notice that in Figure 7.48 all lines have been labelled directly rather than being identified by a separate key. This is strongly recommended wherever possible. Compare the ease with which you can identify the regions with the greatest relative decrease in GDP per head between 1980 and 1989 (North West, North, Yorkshire and Humberside) even using the tangled graphs in Figure 7.48, with the effort needed to interpret Figure 7.46. Even though the chart is in colour, it is difficult to identify the lines correctly. The reader has to scan back and forward to the key to establish that it was Cromwell Road which had NO_2 concentrations above the EC directive limit value. Even without the poorly differentiated colour, the key would still be a barrier to understanding.

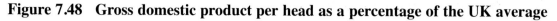

Figure 7.48   Gross domestic product per head as a percentage of the UK average

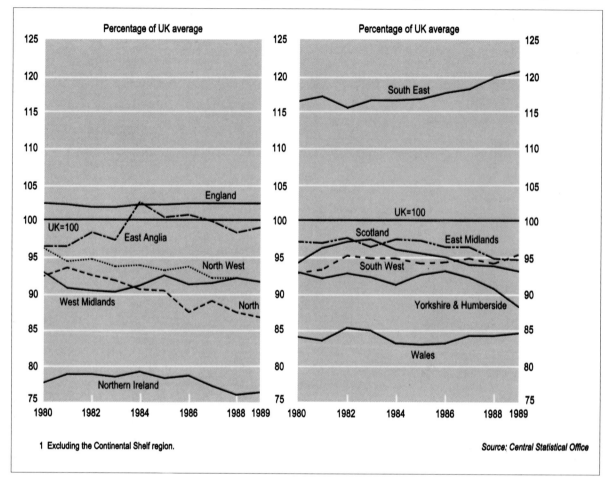

1 Excluding the Continental Shelf region.

Source: Central Statistical Office

Figure 7.49 overpage appears regularly in *Economic Trends*. In this graph there are again too many lines for a detailed message to emerge, but this graph effectively illustrates two features of the data; the general pattern of claimant unemployment has been similar in all regions - falling in 1988 and 1989 and rising during 1991 and 1992 - and that the unemployment rate in Northern Ireland is appreciably higher than in any other region. The chart uses vertical scale lines to help the reader place the movements in time. Such devices can be helpful but, with too many lines already on the chart, the vertical lines add to the cluttered appearance. A final question is whether it is really worth devoting a whole page to a chart with so much data but so little to say.

Figure 7.49   Regional claimant unemployment rates

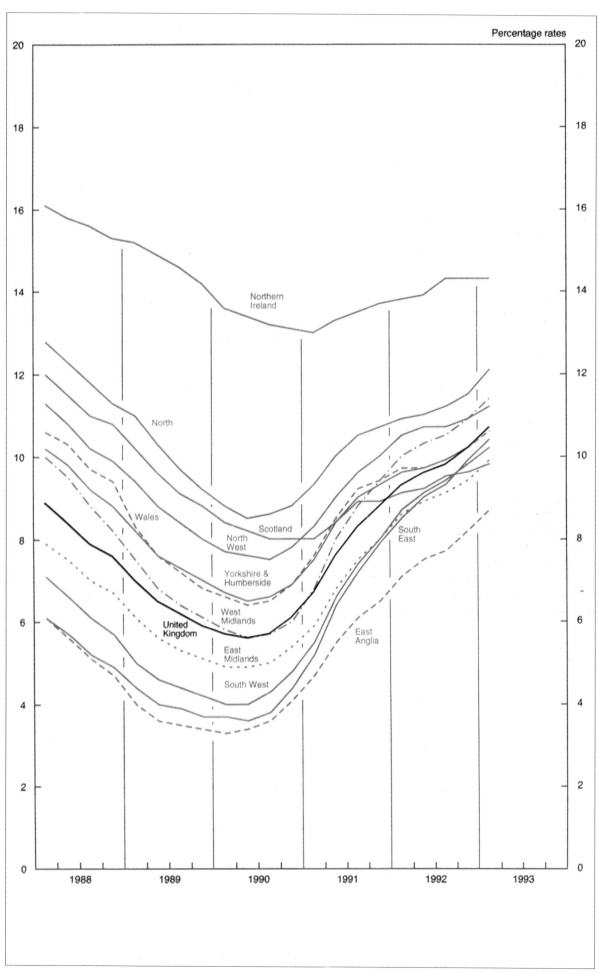

A commonly used variation of line graphs, which cannot be generally recommended, is a surface or layer chart. This sort of diagram serves as an easier-to-draw alternative to the component bar chart and can be effective if kept simple as in Figure 7.50. Too often, the presenter is seduced by the ease with which a lot of information can be included in a single diagram and puts in far too much. What happens then is that, as in Figure 7.51, the reader registers little more than a sort of geological cross-section, and looks in vain for the point being illustrated.

Figure 7.50   Holidays[1] taken by Great Britain residents: by destination

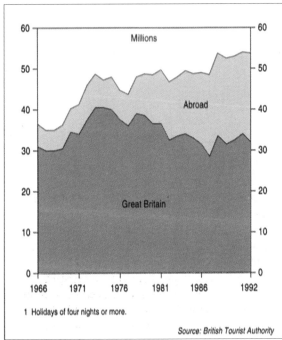

1 Holidays of four nights or more.

Source: British Tourist Authority

Here is the verbal summary which accompanied Figure 7.51 - but is a chart the best way of achieving this?

Verbal summary to Figure 7.51

A breakdown of the capacity of the major generating companies' plant at the end of March each year from 1989 to 1993 is shown in Chart 7.51.

Figure 7.51   Capacity of major power producers

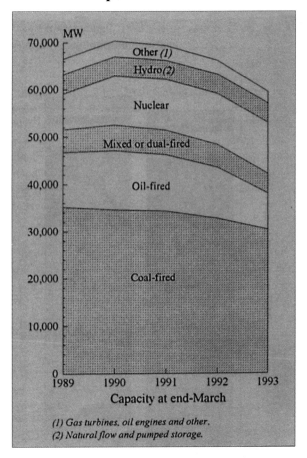

(1) Gas turbines, oil engines and other.
(2) Natural flow and pumped storage.

If a layer chart is to be used the order of the components is important. Can you decide from Figure 7.52 whether there are more 'Commonwealth' and 'other' students in 1991/92 than in 1981/82? After some deliberation you have probably decided there has been little change. Had the order of the components been reversed, 'other' and 'Commonwealth' would have appeared as fairly flat lines leading you directly to this conclusion. Components which change very little are best placed at the bottom of a layer chart.

110

Chart 7.52 Students from abroad: by origin

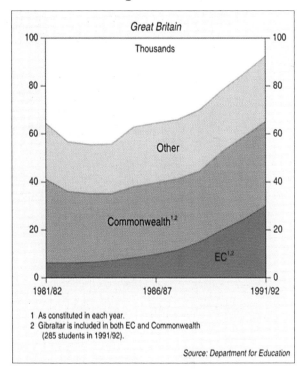

Great Britain

Thousands

Other

Commonwealth[1,2]

EC[1,2]

1981/82   1986/87   1991/92

1 As constituted in each year.
2 Gibraltar is included in both EC and Commonwealth
  (285 students in 1991/92).

Source: Department for Education

Figure 7.53 Estimated and projected[1] total number of children per woman: by woman's year of birth

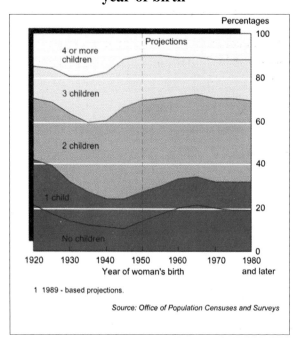

Percentages

Projections

4 or more children

3 children

2 children

1 child

No children

1920   1930   1940   1950   1960   1970   1980
                Year of woman's birth   and later

1 1989 - based projections.

Source: Office of Population Censuses and Surveys

As with bar charts it can be difficult to show movement in the total and changes in the components in a single chart. If the change in the total is illustrated separately, a percentage layer chart can be used to illustrate changes in the components as in Figure 7.53. This chart is not particularly easy to use because of the shape of the data: the lines all change direction several times and it is difficult to see what is happening to the middle groups. However, this kind of data set is difficult to present well and this may be the most effective way of illustrating the basic patterns. Strictly speaking the title should refer to percentages of women to avoid mismatch between the title and the chart. However a good and concise title is difficult for this chart.

Line charts are sometimes shaded for effect as in Figure 7.54. This figure was designed as part of an article celebrating 100 years of the *Employment Gazette* and is clearly intended to be an attractive picture as well as an informative chart. The horizontal scale lines are helpful without being overpowering, but it is difficult to keep track of the timescale when the data points are 3 or 4 inches away from them. If the chart had been drawn to a smaller scale, this problem would have been avoided and some of the small fluctuations would have disappeared, making the broad trend stand out more clearly. Figures 7.42 and 7.43 show equally long timescales in a smaller space.

Figure 7.54   Trade union membership, 1892-1991

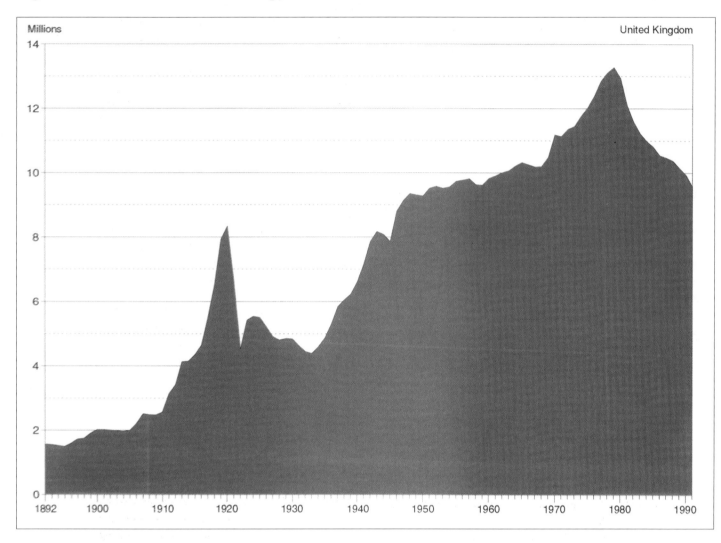

Figure 7.55   Personal sector saving and borrowing

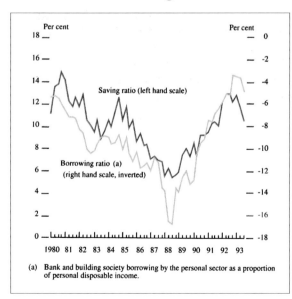

Where two lines are to be compared, they should be drawn on the same scale. Figure 7.55 (with different scales indicated on left and right axes) is difficult to read, even for a specialist audience.

Projections can be distinguished from actual data using a vertical dotted line as in Figure 7.53. Where there are alternative projections or margins of error, the style of Figure 7.56 can be adapted. Sometimes the area of uncertainty can be shaded with or without a central estimate being shown. This effectively reminds readers that forecasts are not 'hard' data.

Figure 7.56   Chlorine loading 1980 to 2020

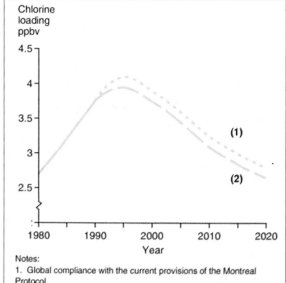

Chlorine
loading
ppbv

Notes:
1. Global compliance with the current provisions of the Montreal Protocol.
2. Global compliance with the current European regulations.

Source: UKSORG

7.4 Scatter diagrams

Scatter diagrams can be used to illustrate a systematic relationship between two measurements (that is, a tendency for an increase in one measurement to be accompanied by a systematic increase or decrease in the second measurement). The chart consists of a horizontal axis calibrated to include the range of one measurement and vertical axis calibrated over the range of the second measurement. Each pair of measurements is then represented by a single point plotted on the chart. For example, Figure 7.57 shows four scatter charts of expenditure on education plotted against GDP per capita. Thus in the top left hand chart Luxembourg is represented by a point plotted at about 16 on the horizontal direction because GDP per head is about $16,000 and just over 5 in the vertical direction because education expenditure per pupil is just over $5,000.

Scatter charts are likely to be less familiar to a general audience than pie charts, bar charts or line graphs. This means that the reader may need some additional help in interpreting the conventions used in drawing the chart and the verbal summary should take account of this.

Where there are a large number of observations and the strength of relationship between measurements is of more interest than the position of individual points, it is not appropriate to label each point.

Figure 7.57 illustrates two measurements which were *positively* correlated with each other, that is, high values of one measurement tended to be associated with high values of the other. The resultant pattern was a scatter of plots rising upwards to the right of the chart. Where two measurements are *negatively* correlated, that is, where high values of one measurement tend to be associated with *low* values of the other, the pattern of points will slope downwards from left to right. Where the measures are closely correlated, all the points will be close to the line. If the points are widely scattered across the chart rather than clustered round the line, there is no simple relationship between two measurements plotted.

Figure 7.57 was accompanied by the verbal summary reproduced under the chart which is unlikely to be fully understood by a non-specialist reader.

7.5 Isotypes

Finally, a method of illustrating changes and comparisons which had considerable popular appeal before the advent of desk top publishing and ready access to colour printing involves the use of isotypes, or pictographs. Here the quantities to be compared are represented by rows or columns of appropriate symbols, for example, simplified silhouettes of men, or ships, or coffee beans. Figure 7.58 shows changes in farm populations in the United States from 1940 to 1973.

The isotype method was developed by Otto Neurath and a team of designers working first in Vienna and later in London over the period 1920 to 1945. ('Isotype' stands for International System of Typographic Picture Education, but the name also implies always using the same symbol.)

Figure 7.57   International comparison of education expenditure per pupil and GDP per head: by level of education establishment

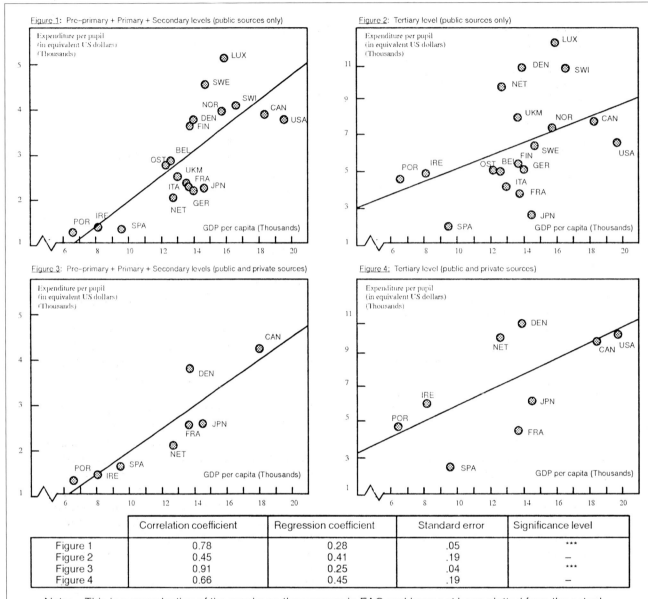

	Correlation coefficient	Regression coefficient	Standard error	Significance level
Figure 1	0.78	0.28	.05	***
Figure 2	0.45	0.41	.19	—
Figure 3	0.91	0.25	.04	***
Figure 4	0.66	0.45	.19	—

Note: This is a reproduction of the graphs as they appear in EAG and have not been plotted from the actual underlying data. Some minor inaccuracies may therefore be present.

Verbal Summary to Figure 7.57

Figure 7.57 shows education expenditure per student by source in relation to per capita GDP and by level of eduction. The indicator places the figures on expenditure per student in relative perspective comparing them against a broad measure of the standard of living of people in each country. One might expect that education expenditure per student of a country would closely correiate with the wealth of its inhabitants (measured here by GDP per Capita). Figures 1 and 3 showing education up to secondary level provides some evidence in support of this, but the tertiary level graphs (figures 2 and 4) do not suggest such a strong relationship at that level (UK data are missing from figures 3 and 4 because of the lack of data available on private expenditure on education).

Figure 7.58   Changes in farm populations, 1940 to 1973

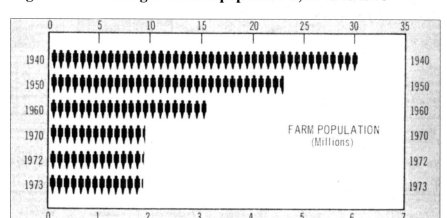

Neurath believed strongly in the power of pictures as communication tools and used the isotype system to illustrate economic and social changes in travelling exhibitions and in books like *Modern Man in the Making*[1]. His view was that 'to remember simplified pictures is better than to forget accurate figures'.

Figure 7.58 is essentially a horizontal bar chart, and was taken from a set of three similar bar charts which illustrated changes in farming in the United States from 1940 to 1973.

Isotypes can also be used to construct grouped bar charts or to illustrate a story. For example children in one line, become adults in a later line.

The basic principles of designing an isotype chart and its basic symbols are explained clearly in the Schmids' book on graphic presentation[2]. It is important that:

1 the symbols should be self-explanatory; thus, if the chart is concerned with ships, the symbol should be the outline of a ship.

2 all the symbols on a chart should represent a definite number of units (for example, 1,000 people). Fractions of this number are represented by fractions of the basic symbol (for example, half a symbol for 500 people).

3 the chart should be made as simple and clear as possible. The number of facts presented should be kept to a minimum.

4 only comparisons should be charted. Isolated facts in themselves cannot be presented effectively by this method.

It is time consuming to produce a clear and professional isotype chart, and a poorly designed or executed one is likely to be inferior to a well executed bar chart. The advantage of using isotypes lies in their ability to 'humanise' the figures. If it is important for your message to appear interesting to the general public and you fear a simple bar chart may not do the job, the use of isotypes should be considered and professional advice sought.

Figures 7.59 and 7.60 are examples of Isotypes from the 1994 issue of *Autocar & Motor* - Figure 7.59 was part of an article discussing the effect of a change in company car taxation. The chart is not specifically referred to in the text, but the key 'break even' points are given succinctly in a small table as shown in the extract given below Figure 7.59. Figure 7.60 is not strictly speaking an isotype, but illustrates the use of pictures to enliven a bar chart. The author clearly believed the reader would find a chart with pictures more interesting on balance than a simple bar chart. The question one needs to ask is whether the pictures add to, or detract from, the chart. The danger is that the pictures merely add clutter and distract the reader from the patterns in the data.

1 NEURATH,O Modern Man in the Making, Knopf, New York 1939

2 SCHMID, CF and SCHMID, SE Handbook of Graphic Presentation, 2nd ed. Wiley, 1979.

Figure 7.59   Taxation of company cars

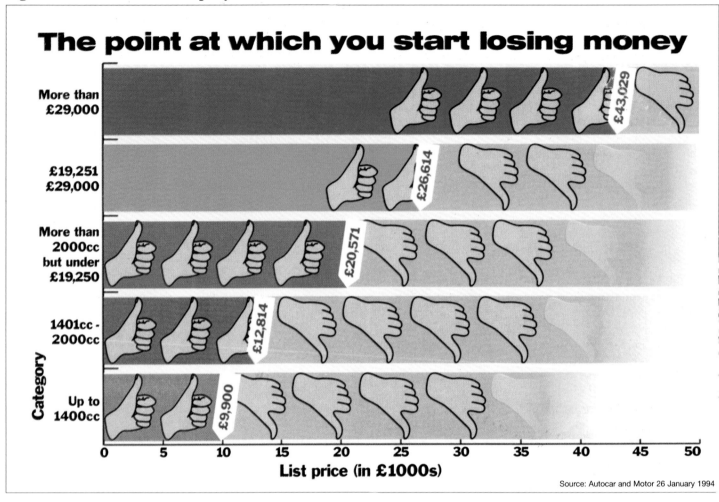

The point at which you start losing money

Category:
- More than £29,000
- £19,251 £29,000
- More than 2000cc but under £19,250
- 1401cc - 2000cc
- Up to 1400cc

£43,029

£26,614

£20,571

£12,814

£9,900

List price (in £1000s)
0   5   10   15   20   25   30   35   40   45   50

Source: Autocar and Motor 26 January 1994

Extract from text accompanying Figure 7.59

The table shows the break-even points, according to list price, in each of the five categories of the previous system. If your car's price falls below the figure quoted, you will be better off under the new system; above it and you'll be worse off.

Category 1	£9,900
Category 2	£12,814
Category 3	£20,571
Category 4	£26,614
Category 5	£43,029

Figure 7.60   Getting away with it

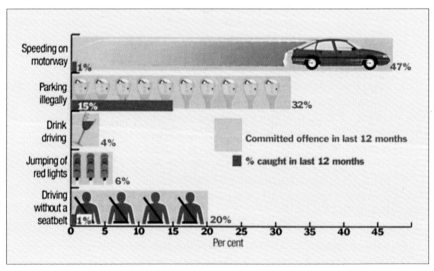

Speeding on motorway — 1% / 47%
Parking illegally — 15% / 32%
Drink driving — 4%
Jumping of red lights — 6%
Driving without a seatbelt — 1% / 20%

Committed offence in last 12 months
% caught in last 12 months

Per cent
0   5   10   15   20   25   30   35   40   45

Source: Autocar and Motor 26 January 1994

7.6 Statistical maps

With the advent of specialised computer packages, statistical maps are increasingly easy to produce and, correspondingly, more frequently encountered. They have three main uses:

- to show location
- to show location and size
- to compare average values for regions or other geographical units.

The first use is simply to show where values of a particular measurement have been recorded within an area, so that clusters or high values can be identified and investigated. For example, plotting every childhood cancer in England on a map would identify any places where there was a cluster of childhood cancers. Statistical maps drawn from space satellites provide a similar level of geographic information. Figure 7.61 for example highlights the high concentrations of lead in topsoils in Wales and the North West of England.

The second use is to record not only the location of certain places but also the value of a variable such as size at each place or site. For example, Figure 7.62 shows both the location of major European ports and their relative sizes in terms of tonnage handled. This chart shows clearly where the ports are, but it is difficult to compare their size despite the key. As noted earlier people find it difficult to compare circles of different sizes accurately. The choice of yellow for the circles is unfortunate because it means the chart cannot be photocopied successfully in black and white.

The third use is to compare measurements of a variable for given geographical units such as regions or counties. For example a map of regions or counties with high and low values of the measurement shown in different colours or shades - as, for example, in Figure 7.63 which shows which areas of the UK have the highest average weekly household disposable income per head.

Figure 7.61   Concentrations of lead in topsoils

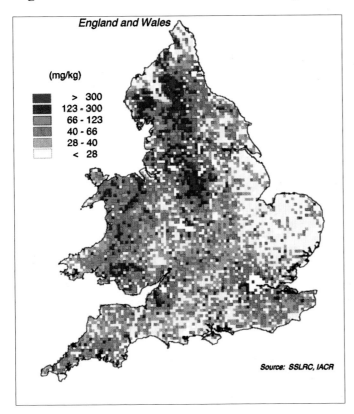

Figure 7.62   Major EC seaports and main inland waterways

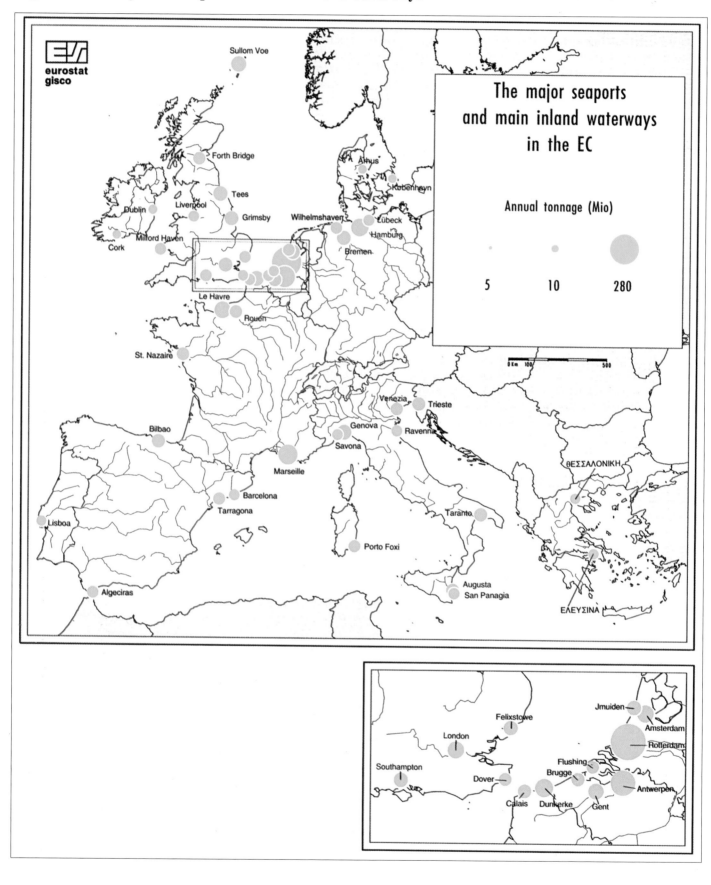

Figure 7.63   Population over retirement age[1], 1991[2]

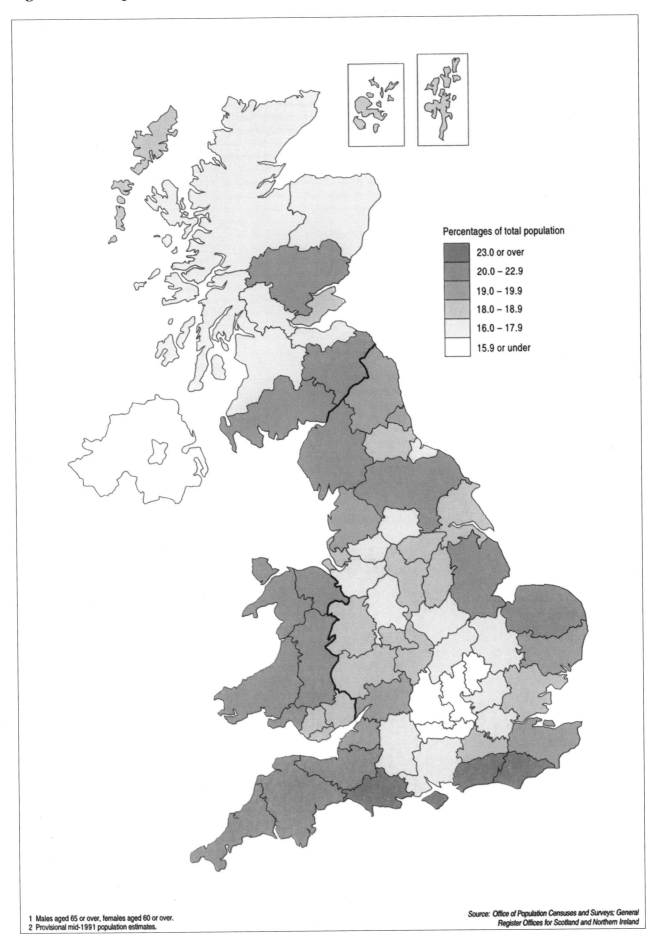

Percentages of total population

- 23.0 or over
- 20.0 – 22.9
- 19.0 – 19.9
- 18.0 – 18.9
- 16.0 – 17.9
- 15.9 or under

1 Males aged 65 or over, females aged 60 or over.
2 Provisional mid-1991 population estimates.

Source: Office of Population Censuses and Surveys; General Register Offices for Scotland and Northern Ireland

In designing the third sort of statistical map it is important to choose the number of categories and the break-points between categories with care. An attractive and comprehensible chart can usually be produced with four or five different categories: Figure 7.64 about lung cancer mortality in Europe, has eight, but would probably have been equally effective with five categories obtained by amalgamating the six middle groups into three broader bands and retaining the two extreme categories. Experience suggests that four shades of a single colour is the most that the eye can distinguish easily and that white should be used to represent the fifth category. With five shades of a single colour it is difficult to see whether areas at opposite ends of the map are in the same shade or not (Figure 7.63). Photocopying into black and white blurs the distinctions still more.

It is important to realise that the selection of each break-point between bands will produce an apparent discontinuity in the measurement charted. Two counties may have very similar percentages of retired residents - say 17.9 per cent and 18.0 per cent - but if the break-point lies between them - say at '18.00 or more' - they will appear as different colours on the statistical map. A good rule is to respect natural breakpoints in the data, rather than using computer default options or an arbitrary selection of regular intervals eg 80 to 99, 100 to 119, 120-139 etc..... If the differences between regions are small, a ranked bar chart will give a clearer picture than a map.

As with all statistical charts the designer has to guard against including so much information on a single chart that no clear message is conveyed - see, for example, Figure 7.65 where groups of bars have been used to show the values of four different measurements in twelve different counties. Readers who are knowledgeable about the subject and familiar with previous versions of this chart will spot significant patterns and changes. However Figure 7.65 is unlikely to work for a general reader unless their attention is drawn to one or two particular features in an accompanying verbal summary; there is just too much data. Similarly a map of the UK showing data at county level contains over 60 pieces of information (one for each area).

Figure 7.64   Lung cancer mortality, 1986

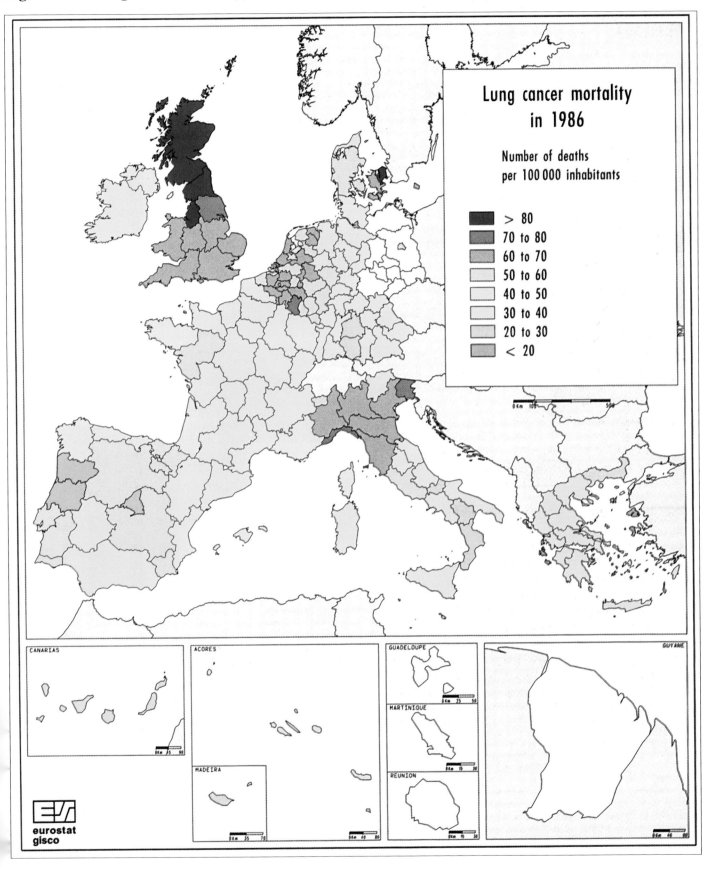

121

Figure 7.65   Housing provision in South East England, 1981-2001

Annual averages

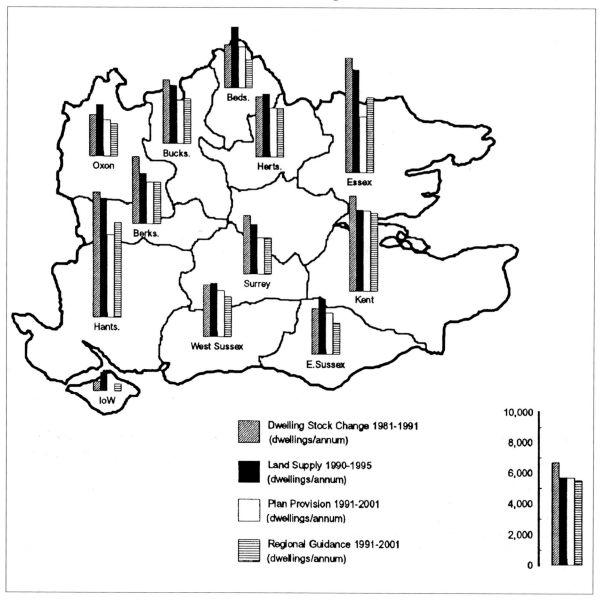

Dwelling Stock Change 1981-1991
(dwellings/annum)

Land Supply 1990-1995
(dwellings/annum)

Plan Provision 1991-2001
(dwellings/annum)

Regional Guidance 1991-2001
(dwellings/annum)

10,000

8,000

6,000

4,000

2,000

0

7.7 Unusual chart forms

The variety of possible chart forms is limited only by the author's imagination. However unusual chart forms should be used with caution. Not only do they require more work on the part of the author, but the reader will have to work harder to understand the chart than if the format were familiar.

Figure 7.66 is admirably simple and uses text as an integral part of the chart.

Figure 7.66   Reduction in number of hares with increasing field size

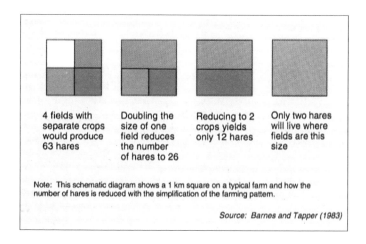

4 fields with separate crops would produce 63 hares

Doubling the size of one field reduces the number of hares to 26

Reducing to 2 crops yields only 12 hares

Only two hares will live where fields are this size

Note: This schematic diagram shows a 1 km square on a typical farm and how the number of hares is reduced with the simplification of the farming pattern.

Source: Barnes and Tapper (1983)

Figure 7.67   Road accident casualties: by hour of the day, 1986

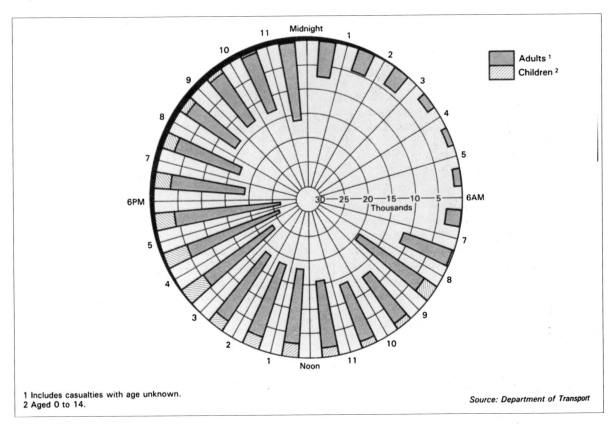

1 Includes casualties with age unknown.
2 Aged 0 to 14.

Source: Department of Transport

Figure 7.67, on the other hand is complicated. It presents data about road accidents at different times of day in the form of a clock. Unfortunately there are 24 hours in a day rather than 12, so the 'clock' has to show 24 hours rather than the 12 that would make it a familiar image. Compare this presentation with Figure 7.68, where the same data are shown as a line graph. Figure 7.67 looks more interesting, but most people find it easier to extract information from Figure 7.68.

Figure 7.68   Road accident casualties on weekdays: by hour of the day, 1992

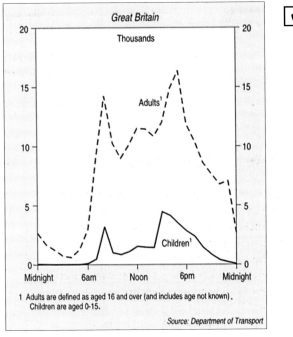

1 Adults are defined as aged 16 and over (and includes age not known).
  Children are aged 0-15.

Source: Department of Transport

123

8 Words

8.1 Don't use words alone

Words alone are very rarely an adequate means of presenting figures. This follows from two well-attested facts:

1 most numbers are meaningless on their own: their significance derives from comparison with other numbers (for example, the latest monthly unemployment figures assume their full significance only when they are compared with, say, the figures recorded for the previous month or those for the same period last year, or when they are assessed jointly with other economic indicators).

2 it is virtually impossible to include more than two or three numbers in narrative without losing clarity.

The main reason for the difficulty of interpreting numbers embedded in prose is that they are scattered: the reader has to scan the text in order to establish the exact definitions of the numbers before he or she can attempt to interpret them. Consider, for example, the following sentence:

Brazil's population has grown at an annual average rate of 2.5 per cent and Malaysia's at 2.6 per cent while real national incomes have increased at annual average rates of 16.5 per cent and 10.5 per cent respectively in the same period.

There are only four numbers in that sentence. It is easy to pick them out but rather less easy to link them up with their labels. To establish that the 10.5 per cent refers to the annual average rate of increase in real national income in Malaysia takes quite a lot of rapid scanning of the lines of text. By contrast, when data are displayed in a table or a chart, the reader knows exactly where to look for the appropriate information. The table or chart heading gives a general statement of what the data represent. Row and column headings in the tables and axis labels in charts give specific definitions of individual numbers. The reader also knows immediately that comparisons can reasonably be made between numbers in the same column or row of a table: physical proximity of numbers in prose is no guarantee that they can be compared directly.

8.2 Vital role of words

However, words have a vital role to play alongside charts and demonstration tables. It is by words that any message is ultimately communicated to its audience. A table or chart without an accompanying verbal commentary is like a silent film: it may be excellently planned and executed, but without words to highlight the important points, indicate general patterns and state the message explicitly, only a few dedicated souls will really understand it.

Guidance in writing lucid prose and constructing memorable reports can be found in other books. For example, the art of writing clear English is well described in Sir Ernest Gowers' *Plain Words*,[1] and *Effective Writing* by Turk and Kirkman[2] deals helpfully with the problems of writing clear reports. The rules and guidelines discussed there apply to writing about figures as well as to writing about, say, the discovery of North Sea gold. The only difference is that while most people would find the discovery of gold under the North Sea intrinsically interesting they find numbers rather boring and alienating. The writer who has to summarise quantitative information must therefore work hard to make his text as readable as possible.

Style guides - such as those produced by *The Times* and *The Economist* - are also helpful in selecting simple alternatives to high sounding phrases and in giving guidance on the use of capital letters and hyphens. A number of government departments now produce their own style guides to encourage uniformly clear writing in all their reports and publications.

---

[1] GOWERS, SIR E *The complete Plain Words*, revised by Sir Bruce Fraser, HMSO, 1973

[2] TURK, C and KIRKMAN, J *Effective Writing* E&FN Spon Ltd, 1982

8.3 Write clearly and concisely

In essence this means that your verbal commentary should be as lucid and as stimulating to the reader as you can make it - while remaining an objective and honest statement of what the numbers say.

Lucidity generally results from an exact understanding of the message you wish to communicate. If you have analysed the data thoroughly and are confident that you know the real patterns and exceptions in the data you will have little difficulty in writing a clear summary. Things to avoid are:

- long sentences, particularly ones with lots of qualifying clauses;

- unfamiliar words: they may be correct; they may be 'stylish', but if there are alternative, everyday words, use them (for example 'alphanumeric data display capability' could be rephrased as 'ability to display letters and numbers');

- technical terms, unless your report will be read only by specialists; (for example correlation coefficient, longitudinal survey);

- sloppy or inexact statements (for example use 'represents' rather than 'relates to' and 'results from' rather than 'reflects');

- too many words: the pithier the prose the sooner it will be read and the better it will be remembered.

It can prove difficult to write exactly without becoming long-winded and it is important to ensure that economy of words does not lead to imprecise statements. Be particularly careful when dealing with changes in percentages and also with proportions of percentages. For example, if you read that 'In 1976, 40 per cent of all applications to industrial tribunals proceeded to hearings while 7 per cent fewer proceeded to hearings in 1992-93', it is not clear whether the percentage of applications which proceeded to hearings in 1992-93 was 33 per cent (40 minus 7) or 37 (which is 93 per cent of 40 per cent). Here a more appropriate ending to the sentence would be, 'while in 1992-93 the corresponding level was 33 per cent.' (The phrase 'percentage points' is sometimes used to clarify these situations but the interpretation of this phrase is not clear to everyone.)

Similarly the sentence 'Between 1978 and 1992 the proportion of applications upheld after proceeding to hearings rose from a quarter to a half' leaves the reader uncertain about whether the quarter and the half refer to a basis of 'all applications' or to basis of 'half those applications which proceeded to hearings'. Since the latter is the correct

interpretation, this summary could be re-phrased as 'In 1978, 25 per cent of applications which proceeded to hearings were upheld. By 1992, the proportion had risen to 50 per cent.'

8.4 Importance of structure and layout

It is helpful to provide as much structure as possible for the reader, particularly in long articles or reports. A helpful and attractive layout also increases the likelihood that the reader will read the whole document attentively or refer back to it later.

Even in a short item you can use bullet points or numbered points as in 'Three major trends can be seen over the last fifteen years. They are:

1 ....
2 ....
3 ....'

This visible structure helps the reader to remember the key points. It also increases the amount of white space on the page which makes the text look less daunting.

In a longer article you can create structure by using plenty of sub-headings. Sub-headings form two important functions: first they tell the reader where he has got to in a report ('Ah yes: now we are comparing the UK figures with the rest of the EU...') and secondly they provide memory cues for the reader when he wishes to recall the content of the report ('And then there was the comparison with the rest of the EU: it said ....).

You can also help the reader by starting your report with a summary of the key points. These 'executive summaries', often in the form of bullet points, are increasingly common.

In general the appearance and accessibility of any statistical report will be improved by:

- generous use of white space (around text not in tables)
- short line lengths
- careful choice of font and font size
- including an executive summary.

8.5 Different sorts of commentaries and different sorts of readers

The best way of making messages interesting to the reader depends upon who the reader is and why he or she is reading the report. In general there are three sorts of statistical commentary:

1 a verbal summary of a data display (usually a single chart or table) which forms one element of a general report covering non-numerical aspects of a topic

2 a covering commentary on a table or set of tables published for future reference

3 a complete report on an essentially statistical investigation.

Each sort of commentary may be read by different sorts of people:

- the general report may be read by a busy executive who will use it as a basis for decision-making, or by someone who want background information on this topic

- the covering commentary may be read by a journalist who is going to write a popular article about the figures or by a specialist or researcher who will use the figures in further analysis

- the report on a quantitative investigation may be read by the person who commissioned the study, by a statistician or scientist interested in the methods used or by someone with a more general interest in the subject of the report.

Before setting pen to paper, be certain you know whom you are writing for:

- the busy executive would like all human knowledge summarised on one sheet of A4 (preferably double spaced) and is most unlikely to read beyond the summary

- the researcher would like a direct and technical explanation of, for example, how the population was stratified and what sort of sampling methods were used

- someone with a general interest in the subject might prefer a journalistic style, with catchy sub-titles and amusing captions.

It is also important to consider the reader's likely attitude to the report and the extent of his or her background knowledge.

If the reader is likely to be:	Then the report should be:
interested and knowledgeable about the subject matter, but not a statistician	concise but non-technical
interested but unfamiliar with the background	a little longer: extra tables and commentary may be needed to set your figures in context
apathetic	thought provoking: journalistic tricks may be useful, such as the use of eye-catching sub-titles; it may help to relate the report to things the reader knows about, for example by using interesting analogies or by referring to well-known events ('This was the last year of the Thatcher administration').
hostile or prejudiced	particularly well-structured and thorough: as well as presenting the main conclusions from an analysis of the data, it may help if you include refutations of likely counter-arguments; but a clear structure must be retained
technically expert (a statistician, an economist or other specialist)	concise and exact in its use of technical terms: but if the report may be of interest to a more general audience, it should start with a non-technical summary.

In sections 8.6 to 8.8 below, we deal with the three main types of commentary. There is inevitably some duplication as the considerations which apply to one type of commentary overlap with those which apply to others.

8.6 Verbal summary of a data display

Most of the attributes of a good verbal summary of a table or chart have already been covered in sections 8.3 to 8.5 and in Chapter 4. The following is a list of good practices to follow in writing such a summary; some of the points listed are discussed at greater length in section 8.8 in the context of writing reports.

1 Keep it short: highlight no more than three or four points and never allow the verbal summary to expand into a blow-by-blow account of each entry in the table or chart.

2 Link the summary closely with the table or chart by quoting numbers directly (as in 'From Chart x we can see that 29 per cent of households had the regular use of one car only in 1961. This proportion increased to 44 per cent in 1971 and then remained more or less constant throughout the 1970s').

3 Avoid emotive or biased descriptions (such as 'shot up by 10 per cent' or 'only rose by a trifling 10 per cent');

4 Unless you are writing specifically for technically expert readers, avoid the use of technical terms (such as 'significant at the 5 per cent level', or 'decreasing marginal rate of return');

5 Do not present changes in definition or other breaks in a series as evidence of trends in the data: the verbal summary should explain what the data reveal of the real situation and should not dwell on the oddities and difficulties of data collection.

Notes on changes in definition or methods of collection should, of course, appear as footnotes and it may sometimes be necessary to refer to such changes in the verbal summary. An introductory sentence along the following lines may then be appropriate. 'From 1991 onwards, Northern Ireland figures are included in the data leading to an apparent sharp increase between 1990 and 1991 in all categories. However, up until 1990 and from 1991 onwards, the most noticeable features of the data....'

Everyday proportions like one in five or two thirds can be used to make a message more accessible to a general audience. Most people are not as comfortable with percentages as the statistician. However it can be confusing if the text says that 'roughly two thirds of people in classes D and E have a bank account' when the figure in the table is 68 per cent. The reader cannot easily locate two thirds in the table.

It can also be confusing if the text quotes figures which are not in the table:- For example quoting a percentage for several groups as a whole when the table includes only the percentages of individual groups.

Examples of clear, concise verbal summaries are given in Chapter 4, pages 31 and 44, and beneath the following table.

Table 8.1, is taken from *Social Trends* 24, and the table itself can be criticised: trends would be more apparent if time were presented vertically and percentages should have been given as whole numbers, but the verbal summary which accompanies it is admirably lucid.

8.7 Commentary on reference tables

Strictly speaking reference tables do not require a commentary: they should be self-explanatory sets of data provided for future use by statistical analysts. However some regularly published reference tables are usually accompanied by a commentary, designed to tell the interested reader about the latest changes in the data. Such a commentary can be provided in either of two ways: each table may be accompanied by its own summary, or a group of tables may be introduced by a general overview of the trends indicated by them all.

Table 8.1 Education and day care of children under five

United Kingdom — Thousands and percentages

		1965/66	1970/71	1975/76	1980/81	1985/86[1]	1989/90	1990/91	1991/92
Children under 5 in schools[2] (thousands)									
Public sector schools									
Nursery schools	- full-time	26	20	20	22	19	17	16	16
	- part-time	9	29	54	67	77	67	68	69
Primary schools	- full-time	209	263	350	281	306	346	357	359
	- part-time	-	38	117	167	228	286	303	318
Non-maintained schools	- full-time	21	19	19	19	20	27	28	29
	- part-time	2	14	12	12	15	19	20	20
Special schools	- full-time	2	2	4	4	4	4	4	4
	- part-time	-	-	1	1	2	2	2	3
Total		269	384	576	573	671	769	799	817
As a percentage of all children aged 3 or 4		*15.0*	*20.5*	*34.5*	*44.3*	*46.7*	*51.3*	*52.8*	*52.8*
Day care places[3] (thousands)									
Local authority day nurseries		21	23	35	32	33	33	33	30
Local authority playgroups					5	5	3	3	2
Registered day nurseries[4]		79	296	401	23	29	64	88	105
Registered playgroups					433	473	491	502	496
Registered child minders[5]		32	90	86	110	157	238	273	297
Total		128	409	522	603	698	830	899	929

1 Data for 1984/85 have been used for Scotland for children under 5 in schools.
2 Pupils aged under 5 at December/January of academic year.
3 Figures for 1965/66 and 1970/71 cover England and Wales at end-December 1966 and end-March 1972 respectively. From 1975/76 data are at end-March except for the Northern Ireland component which is at end-December of the preceding year up to 1987/88.
4 Figures are not available for registered nurseries in Scotland from 1987/88. Estimates have been made for the purpose of obtaining a United Kingdom total.
5 Because of a different method of collection of data relating to registered child minders between 1977/78 and 1980/81, these figures are less reliable. Includes child minders provided by local authorities. English and Welsh data include places for children under 8 in 1991/92.

Source: Department for Education; Department of Health; Welsh Office; The Scottish Office Education Department and Social Work Services Group; Department of Education, Northern Ireland; Department of Health and Social Services, Northern Ireland

Verbal summary to Table 8.1

Education and childcare for children under five

In 1991/92, 53 per cent of all children aged three or four in the United Kingdom attended a school, compared with just 15 per cent in 1965/66 **(Table 8.1)**. This is mainly due to the increase in the number of under fives attending primary schools, both on a full-time and part-time basis. Over the same period, part-time attendance increased eightfold for nursery schools, to 69 thousand in 1991/92, and then ten fold for non-maintained schools, to 20 thousand. Full-time places for the under fives are mainly occupied by four year olds whereas the majority of part-time places are taken by two and three year olds.

In addition to places in schools a growing number of under fives are catered for in day nurseries and playgroups and with child minders. Since 1965/66 there has been a sevenfold increase in such provision. Just over a half of these places were with registered playgroups in 1991/92.

If the first method (a summary for each table) is adopted the commentator's task is comparatively straightforward. All that is needed is a clear, unbiased statement of the latest changes exhibited by the data, and the five points listed in section 8.5 should be observed. Points to bear in mind particularly firmly are:

a it should be easy for the reader to link up the summary statements with the numbers printed in the tables

b emotive descriptions should be avoided (reject words like 'plummeted'; in favour of neutral alternatives like 'fell', 'rose'' and 'remained constant': this is a reference table and obvious impartiality on the part of the presenter is essential).

If a statistical commentary is designed to cover a number of reference tables, as for example the three or four-page commentary which typically accompanies the Labour Market Data section in the middle of the *Employment Gazette*[1], then a slightly different approach is required. Here the commentator's task is to provide an overview of the latest data in a number of related fields: it is reasonable to assume that the reader will be well informed about previous trends but, apart from that, the commentary will have many of the same characteristics of a report on a statistical investigation. Writing such a report is discussed below. An extract from the commentary pages of the *Employment Gazette* is reproduced here as Figure 8.1.

8.8 Report on a statistical investigation

The topic of report writing as a whole is beyond the scope of this book but some points of specific relevance to reports on statistical topics are listed below:

a In writing a report of a statistical investigation, start with the main findings. Every journalist knows that many more people read the first paragraph of any article than read the last. The same applies to statistical reports. By starting with the main findings you ensure that, when the reader is interrupted by the telephone or the need to prepare for a meeting in ten minutes time, at least the most important points have been read. Also, if they were interesting and lucidly written, the reader is more likely to want to read the rest of the report.

b Whenever possible, start a report with clear universal conclusions. You may have to resist a temptation to give first place to some particularly dramatic findings which applied only in limited circumstances. For example, in comparing the fuel consumption of petrol and diesel powered cars, the diesel vehicles might use 30 per cent less fuel in stop-start town driving but only 5 per cent less fuel on steady motorway cruising at 50 miles per hour. Here it would be better to start with the weaker statement 'Under all the driving conditions tested the diesel vehicles consumed at least 5 per cent less fuel than the petrol driven vehicles'. You can then continue the report by explaining that, in some circumstances, the savings in fuel were considerably greater than 5 per cent.

If you start with the dramatic savings of 30 per cent in town driving and then weaken the message by saying that, under other conditions the fuel savings were less noteworthy, the reader instantly suspects that, if he or she reads on far enough, the advantageous performance of the diesel powered cars may vanish entirely.

c The commentator must strive to be objective. This can present problems because the selection of summary statements is inevitably a subjective process. However, if the clearest patterns in the data are the ones selected for comment, then subjective bias will be minimised. By the same token, descriptions should be couched in neutral terms: downward trends should not 'plummet' or 'hurtle'; upward trend should not 'soar' or 'rocket'.

Regrettably, this necessary asceticism can lead to flat and boring prose. The best safeguards against this are a crisp, clear style and a well-structured report.

---

[1] *Employment Gazette:* formerly published monthly by the Employment Department. The section of statistical reference tables is printed on pink paper and contains tables of data on Employment, Unemployment, Vacancies, Industrial Disputes, Earnings, Retail Prices Index and other labour market data.

Figure 8.1 Extract from the Employment Gazette

LABOUR MARKET *update*

Economic background

Table 0.1

❏ **Gross Domestic Product** (GDP) in the fourth quarter of 1993 was 0.8 per cent higher than the previous quarter and 2.6 per cent higher than a year earlier.

❏ **Excluding oil and gas GDP** in the fourth quarter of 1993 was 0.6 per cent higher than the previous quarter and 2.2 per cent higher than a year earlier.

❏ **Retail Sales volumes** in the three months to February were 0.6 per cent higher than in the previous three months and 3.4 per cent higher than a year earlier.

❏ **Manufacturing output** in the three months to January was 0.6 per cent higher than in the previous three months and 2.1 per cent higher than a year earlier.

❏ The **balance of visible trade** in the fourth quarter of 1993 was in deficit by £3.8 billion. This compares to a deficit of £3.0 billion in the previous quarter and £4.2 billion a year earlier.

❏ Excluding oil and erratics **export volumes** in the fourth quarter were 3 per cent lower than the previous quarter and 2½ per cent lower than a year earlier.

❏ Excluding oil and erratics **import volumes** in the fourth quarter were 3½ per cent higher than the previous quarter and 4 per cent higher than a year earlier.

Employment

Figure 1. Tables 1.1 to 1.12, except 1.8

❏ In the quarter to December 1993, the workforce in employment in the UK rose by 15,000, the third successive quarterly rise. *(Table 1.1)*

❏ December's quarterly increase was made up rises of 54,000 self-employed and 22,000 participants on work-related government training schemes, offset by falls of 53,000 employees and 8,000 HM forces. *(Table 1.1)*

❏ Service sector employees fell by 24,000 in the quarter to December following a rise of 162,000 in September - the biggest rise for more than five years. *(Table 1.2)*

❏ Manufacturing employment in Great Britain fell by 13,000 in January following an 11,000 fall in December. *(Table 1.2)*

❏ The January total of 4,163,000 employees in the manufacturing industries is again at its lowest recorded level. *(Table 1.2)*

❏ Overtime worked by operatives rose for the second successive month in January to 9.3 million hours per week.   *(Table 1.11)*

❏ Hours lost through short-time working fell for the second month in January to 0.25 million hours per week. *(Table 1.11)*

Claimant unemployment

Figure 2. Tables 2.1-2.20, except 2.18

❏ UK Seasonally adjusted level of claimant unemployment fell by 38,800 in February to 2,751,800.*(Table 2.1)*

❏ Unemployment level 1,159,400 (73 per cent) higher than in April 1990 when claimant unemployment reached its last trough,

❏ Unemployment level is 207,200 (7 per cent) lower than a year ago.*(Table 2.1)*

❏ The UK seasonally adjusted rate of claimant unemployment, at 9.8 per cent of the workforce, was down 0.1 percentage points on the previous month. Lowest rate since July 1992.*(Table 2.1)*

❏ United Kingdom unemployment rate is 0.7 percentage points lower than 12 months ago and is lower than a year ago in all regions.*(Tables 2.1 & 2.3)*

❏ Between January and February 1994 the level of seasonally adjusted claimant unemployment fell in all regions. The largest percentage falls occurred in East Anglia, the South West and the South East.*(Table 2.3)*

❏ The UK unadjusted total of claimants fell by 47,855 from last month to 2,841,413 or 10.1 per cent of the workforce, an fall of 0.2 percentage points on the rate for the previous month.*(Table 2.1)*

Jobcentre vacancies

Figure 3. Tables 3.1-3.3

❏ The numbers of vacancies remaining unfilled at Jobcentres (UK seasonally adjusted) increased, by 400, to stand at 141,300. This is their highest level since October 1990.*(Table 3.1)*

❏ The seasonally adjusted number of new vacancies notified to Jobcentres increased by 4,200 to 200,800.*(Table 3.1)*

❏ The seasonally adjusted number of people placed into jobs by the Employment Service rose by 2,800, to 150,900, which along with the December 1993 figure is the highest level since April 1990.*(Table 3.1)*

Labour disputes

Figure 4. Tables 4.1, 4.2

❏ It is provisionally estimated that 2,000 working days were lost due to stoppages of work in January 1994. This compares with 1,000 in December 1993 and 49,000 in January 1993.

Figure 1:
Manufacturing and non-manufacturing employees in employment: UK

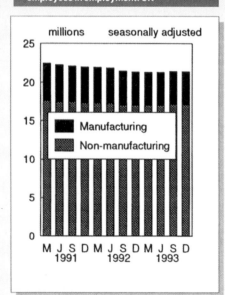

Figure 2:
Unemployment: LFS (ILO definition) and claimant count: GB

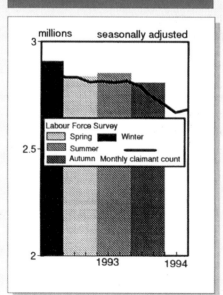

Figure 3:
Unfilled Jobcentre vacancies*: UK

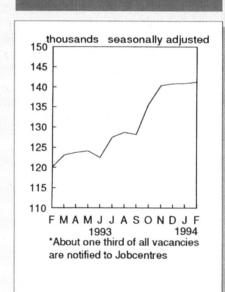

d It must be easy for the reader to link the points discussed in the commentary with the relevant reference tables. This can be achieved in two ways: first by quoting directly from the tables, as in 'The increase over the 12 months to January 1994 in the 'all-items' RPI was 2.5 per cent, up from 1.9 per cent for the 12 months to December. Between December and January the 'all-items' index fell by 0.4 per cent, compared with a fall of 0.9 per cent in January 1993. See (Tables XX or Table XY). (Needless to say, the numbers 2.5, 1.9, 0.4 and 0.9 must all appear in Table XX or Table XY.)

The second way of linking the commentary to the tables is by using sub-headings.

e The reader will find short sections of commentary, each with a distinct sub-heading, much more digestible that a single length of continuous prose. The sub-headings should indicate which tables are commented on in the following section: for example, a heading 'Industrial stoppages' would include comments on tables giving the numbers of industrial disputes and stoppages of work. So long as they maintain the structure of the report clearly, eye-catching subtitles can be extremely effective. Questions, quotations and puns can all be used with the effect of jolting the reader's attention and thus producing landmarks in the report. The following examples are all from the same edition of *The Economist*, a publication which specialises in presenting lucid quantitative analyses: 'Going, going, gone' - above an item on profits in the major British auction houses, 'The nightmare economy' - above charts showing the Ukraine's economic performance in the early 1990s and 'Pricing themselves out of the market' - above a chart on international labour costs. Used skilfully and conservatively captions can enliven the most prosaic report.

f Consider using small additional data displays (charts or tables) where necessary to demonstrate important points: for example, line graphs to illustrate trends over time and well planned demonstration tables to bring together figures from different reference tables. This may also be done if the reference tables are to be set against a wider background, for example, by giving comparable rates for other EU countries

g Do not give explanations about changes in definitions or incompatibilities within the data as part of the commentary: they belong as footnotes.

h Do not put forward changes in the method of data collection as 'reasons' for (observed) patterns. It may be desirable to comment on such a change, usually to prevent the data being misinterpreted - but the comments must be carefully worded. Thus 'The increase in 1992 over 1991 is due to a change in the reporting system' should be rephrased along the lines of 'The new reporting system in 1992 caused reported numbers to increase. It is thought that, under the same reporting system, the 1992 numbers would have been about the same as those recorded in 1991.'

i Avoid technical terms unless you are writing only for a specialist audience. Where a technical term has to be used because of its unique ability to describe an important point, include a non-technical description of its meaning either in the text, if this can be achieved smoothly, or as a footnote, as well as giving a precise definition in the list of definitions at the end of the report. For example, in an article on regional accounts,[1] the following paragraph introduces the non-technical reader to the phrase 'structural component'.

'A question which is of perennial interest in considering the economic problems of the regions is the extent to which overall differences in rates of change are due to the effects of industrial structure on the one hand (the "structural component") or to differences between regional performances within each industry on the other (the "growth component"). An adverse structure would be one which had, for instance, an above average share of typically low-growth industries'.

---

[1] 'Regional Accounts': an article in *Economic Trends* no 349, November 1982, HMSO, pp 82-98

8.9 Examples of good commentary on statistical tables

The final pages of this chapter (Figure 8.2) are taken from a *Home Office Statistical Bulletin* on *Persons granted British citizenship in 1992*. This bulletin is released to the press and public as well as circulated within the department.

The tables and graphs are reproduced in their original form and do not follow all the guidelines suggested in this book (for example the rows and columns of Table 1 are too far apart and the tables are separated from their verbal summaries).

However, the commentary which accompanies them is clear and helpful. The graph has been included because, in general, time trends are easier to see when displayed graphically than from a table, as recommended in section 8.8. The eye-catching front page is a major strength. The four bullet points, summarise key figures and the reader is clearly directed to the paragraphs where these points are amplified. One of the points is illustrated using a simple appropriate chart and the corporate image and GSS logo are included in blue, distinguishing them from the content in black ink.

In reading the commentary in the bulletin note the following virtues:

* the reader is immediately referred to the appropriate table and told that the same data have been displayed graphically

* the summary starts with a clear statement of the broad, overall picture: "42,200 persons were granted British citizenship in 1992, fewer than in 1991 or 1990." This pattern is immediately seen from the graph - but, for closer scrutiny, the numbers can readily be found in Table 1. More detailed comments (about the basis and type of grant) follow next

* some explanation for the fall is offered immediately afterwards in paragraph 1 'The fall partly reflected.....'

* the style of the commentary is crisp and factual: no emotive adjectives or verbs were used

* the report is well structured, with sub-headings used to indicate a change of topic and consideration of a different table

* the important coverage point about Hong Kong is explained clearly in paragraph 2

* no technical terms are used and the patterns in the data are related to events with which the interested reader is likely to be familiar; 'The British Nationality Act of 1981 reduced greatly the number of persons eligible to claim British citizenship by reason of entitlement.'

Figure 8.2   **Extract from Home Office Statistical Bulletin**

Home Office
Statistical Bulletin

Research and Statistics Department
50 Queen Anne's Gate, London SW1H 9AT

SSN 0143 6384

Issue 16/93   24 June 1993

PERSONS GRANTED
BRITISH CITIZENSHIP
UNITED KINGDOM, 1992

MAIN POINTS:

■ 42,200 persons were granted British citizenship in the United Kingdom in 1992, fewer than in 1991 (paragraph 1).

■ 45 per cent were granted on the basis of residence, 35 per cent because of marriage, and 20 per cent were minor children (paragraph 3).

■ Citizens of New Commonwealth countries accounted for over 50 per cent of the total, compared with nearly 80 per cent in 1988 (paragraph 5).

■ The largest nationalities were India (10 per cent), Pakistan (10 per cent) and Bangladesh (5 per cent) (paragraph 5).

**Figure 1   GRANTS OF BRITISH CITIZENSHIP IN THE
UNITED KINGDOM, 1982 - 1992**

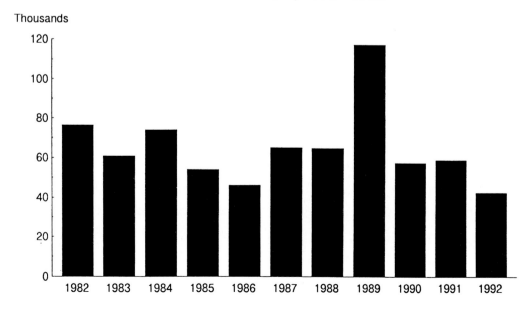

Figure 8.2   Continued

TOTAL GRANTS (Figure 1 and Table 1)

1.   42,200 persons were granted British citizenship in the United Kingdom in 1992, compared with 58,600 in 1991 and 57,300 in 1990.   The fall partly reflected a continuing change in the balance of the work, with the more straightforward applications for registration forming a smaller proportion of the total. In addition, there was a decrease in the number of staff following the closure of the nationality office in Croydon at the end of 1991, when the final stage of the transfer of the work to Liverpool took place. About 36,000 new applications were received in 1992, and the number of applications awaiting decision decreased to around 50,000 at the end of the year.

2.   These data, and those in the rest of the bulletin apart from Table 5, exclude 48,700 persons granted British citizenship in Hong Kong in 1992 under the British Nationality (Hong Kong) Act 1990.

BASIS OF GRANT (Table 1)

3.   As in previous years, a period of residence in the United Kingdom was the most frequent basis on which persons were granted British citizenship in 1992. The number of such grants was 18,200 or nearly 45 per cent of the total, compared with a peak of 70 per cent in 1988-9, and the lowest proportion since 1981. The number of grants because of marriage was 13,900 or nearly 35 per cent of the total, compared with 20 per cent in 1988-9, and the highest proportion for at least ten years. Most (9,300) of the remaining grants were to minor children who accounted for about 20 per cent of the total, a proportion similar to those for most years since 1982.

TYPE OF GRANT (Figure 2 and Table 2) not reprinted in plain Figures

4.   About 85 per cent (35,700) of the grants of British citizenship in 1992 were discretionary, rather than by entitlement. This proportion was the same as that in 1991 and compared with 55 per cent in 1990, 30 per cent on average in 1988-9 and around 40 to 50 per cent in the mid-1980s. The British Nationality Act 1981 reduced greatly the number of persons eligible to claim British citizenship by reason of entitlement. As a consequence many such persons applied in 1987 just before the transitional provisions of the Act ended. The backlog of these 1987 applications was mainly cleared by the beginning of 1991, and therefore subsequently there were relatively few entitlement cases, but more discretionary cases, to be dealt with.

Figure 8.2   Continued

Table 1 Grants of British citizenship[1] in the United Kingdom by basis of grant, 1982-92

Number of persons

Year of grant	All grants	Marriage	Minor children	Residence	Other[2]
1982[1]	76,278	18,641	14,133	43,452	52
1983	60,691	12,191	11,441	31,729	5,330
1984	73,982	22,301	13,826	37,164	691
1985	53,765	15,056	11,034	26,997	678
1986	45,872	12,702	9,447	23,263	460
1987	64,876	16,352	9,502	38,555	467
1988	64,584	13,120	6,423	44,505	536
1989	117,129	22,740	11,830	82,026	533
1990	57,271	15,734	9,534	31,328	675
1991	58,642	19,513	10,646	27,534	949
1992	42,243	13,915	9,346	18,203	779

(1)   Data for 1982 relate to grants of citizenship of the United Kingdom and Colonies.

(2)   Data for 1983 onwards include British Dependent Territories citizens from Gibraltar registered as British citizens under s.5 of the British Nationality Act 1981.

Appendix A: Tables and charts as visual aids

Introduction

The writer of a report is frequently asked to present his or her findings orally. If the report includes statistical analysis, this presentation is likely to require slides, wall charts or viewgraphs (overhead projector slides) of the key charts and tables.

The appendix does not include a technical description of how to prepare slides - many computer packages will now produce these directly. It does, however, offer guidance for those with little experience in the use of visual aids who may not have the appropriate packages and may not have access to the services of a professionally trained graphics officer. Most of the guidelines have been developed from personal experience over a number of years at the Civil Service College and elsewhere, and from both sides of the lecturer's table.

Viewgraphs

The easiest visual aid to prepare and use is the viewgraph or overhead projector slide (OHP slide). The table or chart to be displayed is drawn on an A4 sheet of acetate using either permanent or water-soluble pens, or printed directly onto the acetate using a heat copying machine (but see section 7, below, for possible dangers). For projection, the slide is placed right way up on a projector enabling the speaker to read text from the slide without turning away from the audience. The speaker can point to features of particular interest on the slide using a thin pen or pointer and can add lines or further numbers to the slide during the lecture. There is no need to lower the lights in a lecture room in order to use an overhead projector and, if no screen is available, the image can be projected quite satisfactorily onto a blank wall.

The overhead projector is an extremely valuable, low-tech, lecturing aid. Modern overhead projectors are compact and portable and are automatically provided in many lecture rooms. (But always check in advance.)

Production of viewgraphs

If you plan your presentation well in advance and can call on the services of a graphics officer, your talk can be illustrated with professionally produced viewgraphs. This is obviously ideal. Particularly if you are likely to give the same presentation several times, it is well worthwhile getting a set of expertly executed illustrations.

It is however perfectly possible to produce effective, if amateur, viewgraphs yourself if some simple points are carefully observed. Even when the final slides will be drawn by an expert, it is you, the speaker,

who must plan the content and general design of each slide. A good graphics designer may offer advice about detailed execution, but only the speaker knows exactly what message each slide must contain. The first six points in the following list are therefore applicable both when you intent to draw the slide yourself and when you will leave the production to a professional.

1. *The slide must be readable at the back of the lecture room*. This means:

   - do not put too much on to a single slide: about six lines of well spaced text or numbers will usually be legible
   - use bold letters and numbers
   - use clear colours for numbers and text (never yellow or orange).

2. *It must be intelligible*. This means that it must not contain too much information. One simple message per slide is the general rule, which can, however, be modified by using overlays and reveals to build up a more complex message. (See 3, 4 and 5 below)

3. Overlays can be used effectively to develop an argument. These consist of one or more extra sheets of acetate hinged to the original viewgraph so that they line up exactly with the text or chart on the original slide. They can be used to:

   - superimpose new points, lines or bars on an existing chart
   - produce an extra row or column of figures to supplement a table
   - put summary captions on a slide
   - build up a story caption by caption.

4. Blank space can be left on the slide to be filled in during the talk; for example, the three latest points might be added to a time-trend graph, or a row of average figures might be added to a table of numbers. This approach can be particularly effective if the information added to the chart shows a striking pattern and is likely to produce a livelier, more participative talk than the use of pre-prepared overlays. (Naturally you must have prepared this in advance and be sure that you know exactly what to add and where to write it on the viewgraph: you must also have checked that suitable, working, OHP pens are available during the lecture.) If the slide has been executed in permanent ink and a water-soluble pen is used to fill in extra numbers or lines, additions can be erased later. Alternatively additions can be written on a blank overlay, which prevents any portion of the original slide being accidently erased.

5. Alternatively reveals can be used to build up a story caption by caption. Here all the captions are printed on the same slide which is initially masked so that only a small section is visible to the audience. As a new point is developed in the talk, the mask is progressively removed to reveal the next caption.

6. Use colour constructively. Coloured viewgraphs are always more attractive than black and white ones, but this should not lead to meaningless changes of colour merely for the sake of variety. Different colours can be used in a number of constructive ways. For example, black can be used for tables and diagrams which are reproduced in handouts which accompany the talk, while additional slides or portions of slides are drawn in another colour; in tables, red figures can be used for female numbers and blue figures for male numbers; if summary captions follow a table, the colours of captions can be matched to the colours used for the columns (or rows) of the table; tables can be drawn in one colour, captions in another.

7. *Do not produce viewgraphs by xeroxing standard typing or printing*. Many inexperienced lecturers discover with joy that viewgraphs can be produced by a heat copier from photocopies or from printed tables and charts. This can seem like the ideal way of illustrating a talk with rapidly produced, professional looking slides. It is actually a recipe for producing over-crowded, illegible slides and a restive audience. Standard typeface is too thin and cramped for use on overhead projector slides. If you do not have sufficient time to get slides produced using a special enlarged typeface or on a headliner machine, a slide drawn clearly and lettered by hand, using proper overhead projector pens, will be infinitely better than a slide reproduced from a typed page.

8. If you intend to draw your own slides it is important to lay them out tidily and to make your letters and numbers as regular as possible. The layout should be carefully planned on squared paper (ordinary graph paper can be used) to ensure that columns are vertical and evenly spaced and that rows are regular and horizontal. This paper should then be placed under the acetate when producing the slide.

There are a number of ways of ensuring that letters and numbers are even: one is to use a stencil (which produces a professional-looking end results, but is slow); another is to use squared paper (with a suitable size of square) as a backing sheet and keep each letter or number within a single square; a third way is to write in the gap between two rulers which are kept parallel by sticking them together with sellotape leaving a gap between them equal to the desired letter height.

Use of viewgraphs

Obviously, to use any visual aid effectively, the speaker must have planned his or her talk in detail and have considered exactly when to show each slide.

Three further points are worth noting:

1. When a slide is projected, give the audience time to read it: either remain silent for a minute or two (which can feel like a very long time) or else read it through, quietly but clearly, to the audience. Never talk about something different; no-one can read one message and listen to another.

2. Do not leave a slide on view when you move on to talk about something else. If there is no other slide to be shown, switch the projector off (most projectors are cooled by an obtrusively noisy fan).

3. Finally, obviously - and most importantly - do not block the audience's view of the screen. A sensible precaution is to display your first viewgraph and then ask if everyone can see all of it. If they cannot, the problem can sometimes be solved by projecting the image higher, or by sitting rather than standing to deliver your talk.

35mm slides

The use of good 35mm slides, smoothly projected by remote control at the appropriate point, gives any presentation a thoroughly professional finish.

These slides will always be produced by a professional, though they will be designed by the speaker, and must therefore be planned well in advance. (By contrast, OHP slides can, if necessary, be drawn at home the night before a talk.) The same basic criteria of legibility and intelligibility which were listed in sections 1 and 2 on OHP slides apply here. In summary:

1. Slides must be clear and easy to read: no more than about six lines of text or numbers on a single slide.

2. The must be intelligible: each slide should contain only a single clear message.

3. If a story is to be built up by elaborating on an initial slide, a series of slides must be prepared, each one advancing the story a single step at a time.

35mm slides are loaded into a magazine (or carousel) before the talk in the order in which they will be used and are then projected onto the screen (or blank patch of wall) using a remote control button at the appropriate points during the talk. In general, lights must be dimmed to ensure that the slides are clearly visible.

By comparison with viewgraphs, 35mm slides have two considerable advantages:

- They look more professional

- They are light and compact (several hundred 35mm slides will fit into a chocolate box).

They also have a number of disadvantages (in addition to the need to plan them well in advance):

- It is virtually impossible to alter the order in which they appear: this can be desirable if you have to adapt your presentation in response to unexpected questions from the audience, and it can be achieved relatively smoothly with viewgraphs.

- Similarly it is difficult to recall a slide from the middle of your talk if you wish to review major points at the end.

- It is not possible to use overlays or reveals with 35mm slides (though, of course, a similar effect can be achieved by using a series of slides, each one showing a little more of the story than the previous one).

- You cannot write on 35mm slides.

- You cannot tell by looking at the carousel of slides which slide will appear next. (By contrast, if viewgraphs are store with a sheet of plain paper behind each, you can see the content of the next slide at a glance.)

Wall Charts

Whereas viewgraphs and 35mm slides are normally used to illustrate a sequence of points throughout a talk, each slide appearing before the audience comparatively briefly, wall charts (or large charts supported on an easel) are more appropriate for a table or diagram which will be permanently on display throughout the talk and which will be referred to frequently. Such charts can therefore be used to display information like:

- a map of the geographical area from which data were collected and analysed, subdivided to show regions for which separate results were collected;

- a table showing a list of departments (or firms or areas) which have been studied, along with some basic data relevant to the presentation (for example, the number of staff in each department);

- an outline summary of the presentation showing the structure of the table but omitting any results or details.

Three basic rules apply to the production and use of wall charts.

1. Make sure it is *intelligible*. This involves ensuring that the chart is clearly labelled and also explaining its contents to the audience at an early point in the presentation.

2. Make sure it is *legible*. This means that the chart should be produced using clear colours (not pink, orange, yellow or any pastel shade) and drawn with a thick pen or crayon. The amount of information on the chart must be limited to what can be read easily by the remotest member of the audience. (Clearly this means that more information can be included on a wall chart designed for use in a meeting of six than on one designed for a talk given to an audience of 60.)

3. Make sure it is *visible* without being obtrusive. The chart should be placed where each member of the audience has an uninterrupted view of it and where the speaker can point to it easily. Particularly when other visual aids are also used, wall charts or charts supported on easels should be placed where they will not dominate the presentation. If possible, such charts should be displayed at one side of the speaker's area rather than directly behind the speaker.

Appendix B: Further reading

Demonstration tables
BRITISH STANDARDS INSTITUTION, *Guide to Presentation of Tables and Graphs,* BS7581:1992

EHRENBERG, A S C, *Data Reduction,* (Chapters 1-4)
Wiley, 1975.c

Reference tables
We have not found any general text on this subject. Departmental (or company) guidelines may be available in print: if so they should be considered carefully alongside the recommendations offered in Chapter 5.

Charts
BRITISH STANDARDS INSTITUTION, *Guide to Presentation of Tables and Graphs,* BS7581:1992

CLEVELAND, William S, *The Elements of Graphing Data,* Wadsworth & Broods Kole, 1985

SCHMID, C F and SCHMID, S E, *Handbook of Graphic Presentation,* 2nd ed., Wiley, 1979

SCHMID, CF, *Statistical Graphics: design principles and practices ,* Wiley, 1983

TUFTE, Edward R, *The Visual Display of Quantitative Information,* Graphic Press, 1983

ZELAZNY, Gene, *Choosing and Using Charts,* New York, Video Arts, 1972

Words
GOWERS, Sir E, *The Complete Plain Words,* revised by Greenbaum and Whitcut, HMSO, 1986

HOWARD, Godfrey, *The Good English guide: English usage in the 1990s,* Macmillan, 1993

TURK, C and KIRKMAN J, *Effective Writing,* E&FN Spon, 1982

General style guides
The Economist Style Guide, Economist Publications, 1993

The Times English style and usage, Simon Jenkins, Times Publications, 1992

Word processing style guide: A guide to layout and English usage, Goldingham, TP, Wyndham Press, 1993

Departmental style guides
The writer's guide: Planning, writing and issuing guidance for people in Customs and Excise, Latest edition January 1992

Style guide,
National Audit Office. Latest edition undated but currently available (1995)

Getting your message across: A guide to writing instructions in the ES, Employment Service Communications Service. Latest edition November 1994

Letter writer's kit, The Benefits Agency. Latest edition undated but currently available (1995)

Writing plain English, Inland Revenue. Latest edition undated but currently available (1995)

Straight to the point: A guide to the use of simpler English, Department of Trade and Industry, 1984

Effective written communication: Distance learning package: *Writing good English,* Cabinet Office Training Branch, 1994

Appendix C: Research evidence

This appendix is included in the belief that it is imporant to indicate the reasons for advocating the guidelines offered in the main text. Few of the guidelines are unsubstantiated by evidence that they work well and, in some cases, there is considerable evidence in their favour.

An excellent review article of research in this area was produced by Michael Mcdonald Ross[1], and the interested reader is recommended first to that source.

A brief summary of the research evidence found in a literature review follows, along with references to the original sources.

Appendix D gives precise references to these sources and also to other articles which were consulted in preparing this book.

Summary of research evidence

1. **For reference purposes use tables** (Washburne, 1927; Carter, 1947; Wainer et al, 1980:

   • items are easier to find by searching down a column than along a row (Wright and Fox, 1970)

   • break long columns into blocks of about five items (Tinker, 1960; Wright, 1968; Wright and Fox, 1970)

   • position columns close together (Wright and Fox, 1970).

2. **Do not use text alone** to communicate quantitative information (Washburne, 1927; Feliciano et al, 1963).

3. If the reader is required to register **specific amounts** use **tables** rather than charts (Washburne, 1927; Wainer et al, 1980).

4. **In tables for demonstration** use **heavily rounded** numbers (Washburne, 1927; Ehrenberg, 1977 plus list of associated references).

5. **Arrange tables for demonstration** so as to **highlight patterns** (justification from the field of cognitive psychology, e.g. De Groot, 1966).

6. Except for reference tables, all tables and charts should be accompanied by a **verbal summary** (Feliciano et al, 1963; and 'depth of processing' argument from cognitivie psychologists e.g Craik and Tulving, 1975).

---

[1] MACDONALD ROSS, M. How numbers are shown: a review of research on the presentation of quantitative data in texts. Audio Visual Communication Review, 1977, vol 256, no 4, pp 359-407

Appendix D: Bibliography

American Society of Mechanical Engineers. *Illustrations for Publication and Projection*. National Standards Institute. 1979. (Y15. 1M-1979).

BEEBY, A W and TAYLOR. H P J, How well can we use graphs? *Communicator of Scientific and Technical Information* 1973. vol 17. pp 7-11.

BRINTON, W C. Graphic methods for presenting facts. *New York: The Engineering Magazine Company, 1915.*

BRITISH STANDARDS INSTITUTION *Guide to Presentation of tables and graphs*, BS 7581:1992.

CARTER, L F. An experiment on the design of tables and graphs. *Journal of Applied Psychology*, 1947, vol 31 no 6. pp640-650.

CRAIK, F I M and TULVING, E. Depth of processing and the retention of words in episodic memory. *Journal of Experimental Psychology: General*. 1975, vol 1104, pp 268-294.

CROXTON, F E. Further studies in the graphic use of circles and bars Part 2: some additional data. *Journal of the American Statistical Association,* 1927, vol 22 no 1, pp 36-39.

CROXTON, F E and STEIN, H. Graphic comparison by bars, squares, circles and cubes. *Journal of the American Statistical Association,* 1932, vol 27 no. 1, pp 54-60

CROXTON, F E and STRYKER, R E. Bar charts versus circle diagrams. *Journal of the American Statistical Association*. 1927, vol 22, pp 473-82.

CULBERTSON, H M and POWERS, R D. A study of graph comprehension difficulties. *Audio Visual Communication Review,* 1959, vol 17, pp 97-100.

EELLS, W C. The relative merits of circles and bars for representing component parts. *Journal of the American Statistical Association,* 1926, vol 21 no 2, pp 119-132.

EHRENBERG, A S C. Graphs or Tables? *The Statistician,* 1978, vol 27 no 2, pp 87-96.

EHRENBERG, A S C. Rudiments of numeracy (with discussion). *Journal of the Royal Statistical Society, Series A: General,* 1977, vol 140 part 3, pp277-297.

EHRENBERG, A S C. 'What we can and can't get from graphs, and why.' (Paper based on invited talk at Statistical Meetings. Detroit, August 1981). 1982.

FEINBERG, S E. Graphical methods in statistics. *The American Statistician,* 1979, vol 33 no4, pp 165-178.

FELICIANCO, G D, POWERS, R D and KEARL, B E. The presentation of statistical information. *Audio Visual Communication Review,* 1963. vol 11, pp 33-39.

FLANNERY, J J. The relative effectiveness of some common graduated point symbols in the representation of quantitative data. *Canadian Cartographer,* 1971, vol 8 pp 96-109.

FLETCHER, J. How to Write a Report. Institute of Personnel Management, 1983.

GOWERS, SIR E. The Complete Plain Words, revised by Greenbaum and Whitcut, HMSO, 1986.

GREGG, V. *Human Memory.* Methuen, 1975.

DE GROOT, A D. Perception and memory versus thought: some old ideas and recent findings. *In* B Kleinmuntz (ed). *Problem Solving,* 1966.

GROPPER, G L. Why IS a picture worth a thousand words? *Audio Visual Communication Review,* 1963, vol 11, pp 75-95.

HABER, R N. How we remember what we see. *Scientific American,* 1970, vol 222 part 2, pp 104-112.

HAMMERTON, M. How much is a large part? *Applied Ergonomics,* 1976, vol 7 no 1, pp 10-12.

HARTLEY, J and BURNHILL, P. Fifty guidlines for improving instructional text. *Programmed Learning and Education Technology,* 1977, vol 14 no 1, pp 65-73.

HYDE, J S and JENKINS, J J. Recall of words as a function of semantic, graphic and syntactic orienting tasks. *Journal of Verbal Learning and Verbal Behaviour,* 1973, vol 12 no 1, pp 147-180.

MACDONALD ROSS, M. How numbers are shown: a review of research on the presentation of quantitative data in texts. *Audio Visual Communication Review,* 1977, vol 25 no 4, pp 359-407.

MACDONALD ROSS, M. Research in graphics communication: graphics in text, how numbers are shown. 1977. *Institute of Educational Technology, Monograph no 7*, Open University.

MILLER, G A. The magical number seven, plus or minus two. *Psychological Review,* 1956, vol 63 no 2 pp 81-97.

MEIFOEFER, H J. The visual perception of the circle in thematic maps: experimental results. *Canadian Cartographer,* 1973, vol 10, pp 63-84.

NEURATH, M. Isotype. *Instructional Science,* 1974, vol 3 no 2, pp 127-150.

NEURATH, O. International Picture Language, Kegan Paul, Trench and Trubner, 1936.

NEURATH, O. Modern man in the making. Knopf. New York, 1939.

NICKERSON, R S. Short-term memory for complex meaningful visual configurations: a demonstration of capacity. *Canadian Journal of Psychology,* 11965, vol 19 no 2 pp 155-160.

PAIVIO, A and CSAPO, K. Concrete image and verbal memory codes. *Journal of Exptl. Psychology,* 1989, vol 80 no 2, pp 279-285.

PEARSON, E S. Some aspects of the geometry of statistics. *Journal of the Royal Statistical Society, Series A: General.* 1956a, vol 119 no 2, pp 125-149.

PEARSON, L V and SCHRAMM, W. How accurately are different kinds of graphs read? *Audio Visual Communication Review,* 1955, vol 2 no 2, pp 178-189.

PLAYFAIR, W. The Commercial and Political Atlas, 3rd ed. London: J Wallis, 1801.

SCHUTZ, H G. An evaluation of formats for graphic trend displays (Parts A and B). *Human Factors*, 1961, vol 3, pp 99-119.

SCHMID, C F. Statistical Graphics: design principles and practices. Wiley, 1983.

SCHMID, C F and SCHMID, S E. Handbook of Graphic Presentation, 2nd ed. Wiley, 1979.

SHEPARD, R N. Recognition memory for words, sentences and pictures. *Journal of Verbal Learning and Verbal Behaviour,* 1967, vol 6 no 1, pp 156-163.

SIMON, H A. Models of Thought. Yale University Press, 1979.

SIMON, H A and GREGG, L W. One trial and incremental learning. Chapter 3.3 of *Models of Thought* by H A Simon. Yale University Press, 1967.

TINKER, M A. Legibility of mathematical tables. *Journal of Applied Psychology,* 1960, vol 44 no2, pp 83-87.

TUKEY, J W. *Exploratory Data Analysis.* Addison-Wesley, 1977.

TULVING, E. Subjective organisation and effects of repetition in multi-trial free-recall learning. *Journal of Verbal Learning and Verbal Behaviour,* 1966 vol 5 no 2, pp 193-197.

STANDING, L. CONEZIO, J AND HABER, R N. Perception and memory for pictures: single trial learning of 2560 stimuli. *Psychonomic Science*, 1970 vol 19 pp 73-74.

VERNON, M D. Learning from graphical material. *British Journal of Psychology,* 1946, vol 36 pp 145-158.

VERNON, M D. The use and value of graphical material presenting quantitative data. *Occupational Psychology* 1952, vol 26 no 1, pp 22-34 and no 2, pp 96-100.

VERNON, M D. Presenting information in diagrams. *Audio Visual Communication Review,* 1953. vol xx pp 147-158.

WAINER, H, LONO, M and GROVES, C. On the display of data: some empirical findings. 1980. Washington DC. *Bureau of Social Science Research. Draft report.*

WASHBURNE, J N. An experimental study on various graphic, tabular and textual methods of presenting quantitative material. *Journal of Educational Psychology* 1927, vol 18 no 6 pp 3361-376 and no 7, pp 465-476.

WAUGH, N C. Retrieval time in short-term memory. *British Journal of Psychology,* 1970 vol 6 no x pp 1-12.

WAUGH, N C and NORMAN, D A . Primary memory *Psychological Review,* 1965, vol 72 no. 2, pp 89-104.

WEINTRAUB, S. What research says to the reading teacher (graphs, charts and diagrams). *Reading Teacher.* 1967, vol 20 pp 345-349.

WRIGHT, P. Using tabulated information. *Ergonomics.* 1968, vol 11 no 4, pp 331-343.

WRIGHT, P and FOX, K. Presenting information in tables. *Applied Ergonomics,* 1970, vol 1 no 4 pp 234-242

WRIGHTSTON, J W. Conventional versus pictorial graphics. *Progressive Education, 1936, vol 13 no 6, pp 460-462.*

ZELAZNY, G. Choosing and Using Charts. *London. Video Arts,* 1972.

Appendix E: Sources of tables and charts

This index gives the publication from which the tables and charts have been reproduced. Some tables were prepared specially for Plain Figures by the authors; these tables are labelled "Original".

Chapter 6

Chapter 7

Chapter 8

Printed in the U.K for *The* Stationery Office
Dd: 0302179 12/96 56-9308 77984